THIRD EDITION

The Basics of

Oil Spill Cleanup

THIRD EDITION

The Basics of

Oil Spill
Cleanup

Merv Fingas

CRC Press
Taylor & Francis Group
Boca Raton London New York

CRC Press is an imprint of the
Taylor & Francis Group, an **informa** business

CRC Press
Taylor & Francis Group
6000 Broken Sound Parkway NW, Suite 300
Boca Raton, FL 33487-2742

© 2013 by Taylor & Francis Group, LLC
CRC Press is an imprint of Taylor & Francis Group, an Informa business

No claim to original U.S. Government works

Printed in the United States of America on acid-free paper
Version Date: 20121030

International Standard Book Number: 978-1-4398-6246-9 (Hardback)

Library of Congress Cataloging-in-Publication Data

Fingas, Mervin F.
　　The basics of oil spill cleanup / Merv Fingas. -- 3rd ed.
　　　　p. cm.
　　Summary: "An examination of pollution caused by crude oils and petroleum products derived from them, this book covers how oil spills are measured and detected and discusses the properties of the oil as well as its long-term fate in the environment. This third edition contains a new chapter devoted to pollution effects on wildlife. It focuses on the cleanup of oil spills that occur in water, since these spills spread most rapidly and cause the most visible environmental damage. It also includes coverage of the latest technologies as well as recent spills, including the Gulf of Mexico"-- Provided by publisher.
　　Includes bibliographical references and index.
　　ISBN 978-1-4398-6246-9 (hardback)
　　1. Oil spills--Cleanup. 2. Oil pollution of the sea--Environmental aspects--North America. I. Title.

TD427.P4F55 2012
628.1'6833--dc23
2012032425

Visit the Taylor & Francis Web site at
http://www.taylorandfrancis.com

and the CRC Press Web site at
http://www.crcpress.com

Contents

Preface

This book is a revised and expanded edition of *The Basics of Oil Spill Cleanup*, which was first published by Environment Canada in 1978 and subsequently revised in 2000 by the present author. With the rapid progress in cleanup technology since that time, another edition was certainly needed. This third edition is designed to provide a broad knowledge of the cleanup and control of oil spills. It is aimed at both the general public and those who actually deal with the cleanup of oil spills, although it is not intended to serve as a field manual. The cleanup of oil spills that occur on water is emphasized since these spills spread most rapidly and cause the most visible environmental damage.

The book deals primarily with crude oils and petroleum products derived from crude oils. In addition to cleanup techniques, the book covers how oil spills are measured and detected, and the properties of the oil and its long-term fate in the environment. The effects of an oil spill on the environment and the effectiveness of cleanup and control vary significantly with the type of oil spilled. The types of oil are reviewed to help the reader understand the different cleanup and control measures needed for different types of oil and environmental circumstances. The final chapter of the book covers the effects of oil spills on the environment and wildlife. As the effects of oil on the environment are serious and determine how and why we clean up spills, a summary of this topic seems an appropriate way to end this volume. A list of further reading is provided. A glossary of technical terms is provided at the end of the book.

Acknowledgments

Among those who provided photographs and other information for this publication are Al Allen of Spiltec in Woodinville, Washington; Fred Beech, Paul Parete, and Ali Khelifa, of Environment Canada; ITOPF (International Tanker Owners Pollution Federation) in London; Jeremy Robida of PWS RCAC in Anchorage, Alaska; Peter Lane and Peter Newson of Applied Fabric Technologies, Orchard Park, New York; Stuart Ellis of Elastec; Victor Bennett of Aquaguard, Vancouver, British Columbia, Canada; Sergio Difranco of the Canadian Coast Guard; Jacqui Michel of RPI; Ed Stanton; Robert Pond of the United States Coast Guard; and Steve Ricks of the Marine Spill Response Corporation (MSRC).

Many others, too numerous to list here, provided comments, support, advice, and encouragement. The author gratefully acknowledges their contributions.

Author

Merv Fingas, M.Sc., MBA, Ph.D., has worked for more than 38 years in the field of oil spill technology. He was chief of Environment Canada's Environmental Emergencies Technology Centre in Ottawa for more than 30 years. Over the years, he has conducted research in spill dynamics and behavior; studies of spill treating agents; in-situ burning of oil; and oil spill chemistry and analysis.

Dr. Fingas earned his doctorate in environmental sciences from McGill University in Montreal, Quebec. He also holds master's degrees in science and in business from the University of Ottawa in Ontario; and a bachelor degree of arts and technical training both in machining and in electronics.

Dr. Fingas has authored or coauthored more than 800 technical reports and papers on various aspects of oil or chemical research. These include topics such as oil emulsion formation, oil evaporation, treating agent testing and use, studies of oil analysis and fingerprinting, in-situ burning of oil, oil spill remote sensing, and personal protection equipment. This will be his seventh book.

In addition to being a reviewer for numerous journals, Dr. Fingas is also on many editorial boards for scientific journals including the *Journal of Hazardous Materials*, the *Journal of Micro-Column Separations*, and the *International Journal of BioSciences and Technology*. He was an editor of the *Journal of Hazardous Materials* for a 6-year term, which recently ended. He is chairman of the in-situ oil spill burning subcommittee of ASTM F-20 on hazardous materials and chairman of the treating agent subcommittee and the remote sensing subcommittee of the same organization. Dr. Fingas has been a member of several United States Academy of Science committees on oil spill topics.

List of Illustrations

List of Tables

Introduction

Chapter 1, "Oil Spills: Why Do They Happen and How Often?" deals with the questions of why oil spills happen and provides statistics on how often and where they occur. This includes a summary of American and Canadian spills, sources of oil spills into the seas worldwide, and a list of the largest oil spills that have occurred in the last 40 years. Chapter 2 provides information on planning the response to an oil spill and the functions of oil spill response organizations in industry and government. Topics covered include contingency plans, training, the structure of response organizations, the duties of the on-scene commander and response team, oil spill cooperatives, and the role of other organizations and contractors in oil spill cleanup.

The chemical composition and physical properties of the different types of oil are described in Chapter 3. The oils that are used in this book illustrate the fate, behavior, and cleanup of oil spills. These represent the primary oil and petroleum products used and spilled. They are gasoline; diesel fuel; a light crude oil; a heavy crude oil; an intermediate fuel oil (IFO), which is made from a heavy residual oil and diesel fuel fraction; a residual oil, sometimes called Bunker fuel; and a crude oil emulsion.

When oil spills on water, various transformation processes occur, which are referred to as the "behavior" of the oil. Two types of transformation processes are discussed in Chapter 4. The first is weathering, with emphasis on evaporation, the formation of water-in-oil emulsions, and natural dispersion. The second is a group of processes related to the movement of oil in the environment. Spill modeling, whereby the behavior and movement components of an oil spill are simulated using a computerized model, is discussed.

Chapter 5 reviews the technologies used to detect and track oil slicks. This includes both surface techniques and remote sensing techniques from aircraft and satellites, which are especially useful when oil is difficult to detect, such as at night, in ice, or among weeds. The analysis of samples to determine the oil's properties, its degree of weathering, its source, and its potential impact on the environment is discussed.

The most common way to contain oil on water is to use devices known as booms. Chapter 6 explores the types of booms, their construction, operating principle and uses, as well as their limitations. It also covers ancillary equipment used with booms, sorbent booms, and special purpose booms. Chapter 7 summarizes methods to physically recover oil from the water surface, usually after it has been contained using booms. Devices known as skimmers are available to recover oil. The effectiveness and advantages and disadvantages of various types of skimmers are presented. The use of sorbents, material that absorbs the oil, is reviewed. In some cases, the oil is recovered manually and often all of these approaches are used in a spill situation.

Each method has limitations, depending on the amount of oil spilled, sea and weather conditions, and the geographical location of the spill.

Storage, separation of oil from water and debris, and disposal of the oil are crucial parts of a cleanup operation. Chapter 8 describes temporary storage, separation, and disposal, as well as the types of pumps used to move the oil. Treating the oil with chemical agents is another option for cleaning up oil spills on water. The use of these agents is discussed in Chapter 9. Dispersants are agents that promote the formation of small droplets of oil that disperse throughout the water column. Their effectiveness, toxicity, and application are reviewed. Other agents discussed are surface-washing agents or beach cleaners, emulsion breakers and inhibitors, recovery enhancers, solidifiers, sinking agents, and biodegradation agents.

In-situ burning is an oil spill cleanup technique that involves controlled burning of the oil at the spill site. The advantages and disadvantages of this technique are highlighted in Chapter 10, as well as conditions necessary for igniting and burning oil, burning efficiency and rates, and how containment is used to assist in burning the oil and ensure that the oil burns safely. The air emissions produced by burning oil are described and the results of the analytical studies into these emissions are summarized.

Oil spills on shorelines are more difficult and time-consuming to clean than spills in other locations and cleanup efforts on shorelines can cause more ecological and physical damage than if the removal of the oil is left to natural processes. Chapter 11 features the important criteria that are evaluated before deciding to clean oil-contaminated shorelines. These criteria include the behavior of oil in shoreline regions, the types of shorelines and their sensitivity to oil spills, the assessment process, shoreline protection measures, and recommended cleanup techniques.

Although oil spills on land are easier to deal with and receive less media attention than spills on water, oil spills on land make up the vast majority of oil spills in Canada. Chapter 12 describes the varying effects and behavior of oil on different habitats and ecosystems. Spills that occur primarily on the surface of the land and those that occur partially or totally in the subsurface, and the different containment and cleanup methods for each type of spill are outlined.

Chapter 13, "Effects of Oil Spills on the Environment," reviews the many and varied effects of oil on different elements of the environment and summarizes the state of the art in assessing the damage caused by oil spills. The effects of oil on various organisms in the sea are discussed, as well as effects on freshwater systems, on land biota, and on birds.

1

Oil Spills: Why Do They Happen and How Often?

Major oil spills attract the attention of the public and the media. In recent years, this attention has created a global awareness of the risks of oil spills and the damage they do to the environment. Oil is a necessity in our industrial society, however, and a major component of our lifestyle. Most of the oil and petroleum products used in Canada and the United States are for transportation. According to trends in petroleum usage, this is not likely to decrease much in the future. Industry uses oil and petroleum derivatives to manufacture such vital products as plastics, fertilizers, and chemical feedstocks, which will still be required in the future.

In fact, the production and consumption of oil and petroleum products are increasing worldwide, and the threat of oil pollution is increasing accordingly. The movement of petroleum from the oil fields to the consumer involves as many as 10 to 15 transfers between many different modes of transportation including tankers, pipelines, railcars, and tank trucks. Oil is stored at transfer points and at terminals and refineries along the route. Accidents can happen during any of these transportation steps or storage times.

Obviously, an important part of protecting the environment is ensuring that there are as few spills as possible. Both government and industry are working to reduce the risk of oil spills with the introduction of strict new legislation and stringent operating codes. Industry has invoked many operating and maintenance procedures to reduce accidents that lead to spills. In fact, the rate of spillage has decreased in the past 10 years. This is especially true for tanker accidents. Intensive training programs have been developed to reduce the potential for human error. Despite this, spill experts estimate that 30% to 50% of oil spills are either directly or indirectly caused by human error, with 20% to 40% of all spills caused by equipment failure or malfunction.

There are also many deterrents to oil spills including government fines and the high cost of cleanup. In Canada, it costs an average of $50 to clean up each liter of oil spilled. In the United States, these costs average about $200 per liter spilled. The average cost of cleanup worldwide ranges from $40 to $400 per liter, depending on the type of oil and where it is spilled. Cleaning up oil on shorelines is usually the most expensive cleanup process.

PHOTO 1.1
The 1991 Arabian Gulf oil spill was the worst on record. This photo shows a portion of Saudi shoreline that was heavily oiled.

How Often Do Spills Occur?

Oil spills are a frequent occurrence, particularly because of the heavy use of oil and petroleum products in our daily lives. About 450,000 tons of oil and petroleum products are used in Canada every day. The United States uses about 10 times this amount and worldwide about 20 million tons are used per day. Most domestic oil production in Canada is from more than 200,000 oil wells in Alberta and Saskatchewan. Increasingly, oil also comes from the Alberta oil sands in the form of bitumen. There are 20 oil refineries in Canada, 5 of which are classified as large. Canada imports little crude oil or other products per day but exports about 500,000 tons per day, mostly to the United States.

In the United States, more than half of the approximately 4 million tons of oil and petroleum products used per day is imported, primarily from Canada, Saudi Arabia, and Africa. About 40% of the daily demand in the United States is for automotive gasoline and about 15% is for diesel fuel used in transportation. About 40% of the energy used in the United States comes from petroleum, 25% from natural gas, and 20% from coal.

In both Canada and the United States, much of the refined oil goes into powering transportation. Figure 1.1 shows the typical outputs from a refinery in terms of percent of products. One might expect that these products might be spilled in similar amounts. However, many of the products are handled differently, resulting in different spill percentages.

FIGURE 1.1
The typical output of a refinery from one input unit of crude oil.

Spill statistics are collected by a number of agencies in Canada and the United States. In Canada, provincial offices collect data and Environment Canada maintains a database of spills. In the United States, the Coast Guard maintains a database of spills into navigable waters, while state agencies keep statistics on spills on land, which are sometimes gathered into national statistics. The Bureau of Safety and Environmental Enforcement (BSEE) in the United States maintains records of spills from offshore exploration and production activities.

It can sometimes be misleading to compare oil spill statistics, however, because different methods are used to collect the data. In general, statistics on oil spills are difficult to obtain and any data set should be viewed with caution. The spill volume or amount is the most difficult to determine or estimate. For example, in the case of a vessel accident, the exact volume in a given compartment may be known before the accident, but the remaining oil may have been transferred to other ships immediately after the accident. Some spill accident data banks do not include the amounts burned, if and when that occurs, whereas others include all the oil lost by whatever means. Sometimes the exact character or physical properties of the oil lost are not known and this leads to different estimations of the amount lost.

Reporting procedures vary in different jurisdictions and organizations, such as government or private companies. Minimum spill amounts must be reported according to various regulations, depending on the product spilled. Spill statistics compiled in the past are less reliable than more recent data because few agencies or individuals collected spill statistics before about 1975. Nowadays, techniques for collecting statistics are continually improving.

The number of spills reported also depends on the minimum size or volume of the spill. In both Canada and the United States, most oil spills reported are more than 4000 L (about 1000 gallons). In Canada, there are about 12 such oil spills every day, of which only about one is spilled into navigable waters. These 12 spills amount to about 40 tons of oil or petroleum product. In the United States, there are about 15 spills per day into navigable waters and an estimated 85 spills on land or into freshwater.

Despite the large number of spills, only a small percentage of oil used in the world is actually spilled. Oil spills in Canada and the United States are summarized in Figures 1.2 and 1.3 in terms of the volume of oil spilled and the actual number of spills. In terms of oil spills, it can be seen from these figures that there are differences between the two countries.

There are more spills from barges into navigable waters in the United States proportionately than in Canada because more oil is transported by barges. In fact, the largest volume of oil spilled in water in the United States comes from barges, whereas the largest number of spills comes from vessels other than tankers, bulk carriers, or freighters.

In Canada and the United States, most spills take place on land and this accounts for a high volume of oil spilled. Pipeline spills account for the highest volume of oil spilled. In terms of the actual number of spills, most oil spills happen at petroleum production facilities, wells, production collection facilities, and battery sites. On water, the greatest volume of oil spilled comes from marine or refinery terminals, although the largest number of spills is from the same source as in the United States, that is, vessels other than tankers, bulk carriers, or freighters.

The sources of oil released into the sea are shown in Figure 1.4. More than half of the oil in the sea derives from natural sources include the many natural "seeps" or discharges from oil-bearing strata on the ocean floor. About 38% of the oil that reaches the seas is the runoff of oil and fuel from land-based sources, usually from wastewater. Significant amounts of lubricating oil finds its way into wastewater that is often discharged directly into the sea. About 13% of oil reaching the sea comes from the transportation sector, which includes tankers, freighters, barges, and other vessels.

A list of the largest oil spills in the last 40 years is provided in Table 1.1. Data are derived from the author's records and the general literature on oil spills. The spills are listed according to their volume beginning with the largest spill to date: the release of oil during the Gulf War in 1991. There have been several large oil spills from pipelines, storage tanks, and blowouts at production wells. It is important to note that because of the differences in record keeping and spill reporting, the list of the largest spills may vary from source to source.

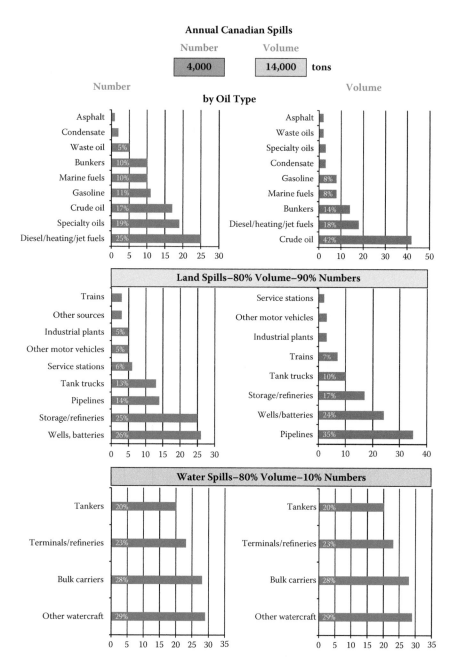

FIGURE 1.2
Typical spill statistics for Canada in a given year.

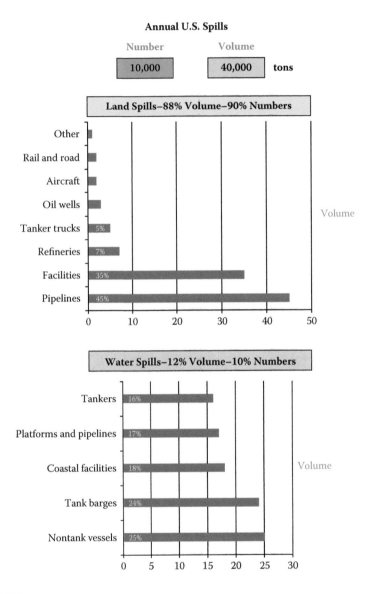

Annual U.S. Spills

FIGURE 1.3
The typical spill statistics for the United States in a given year.

The public often has the misconception that oil spills from tankers are the primary source of oil pollution in the marine environment. Although it is true that some of the large spills are from tankers, it must be recognized that these spills still make up less than about 5% of all oil pollution entering the sea. The sheer volume of oil spilled from tankers and

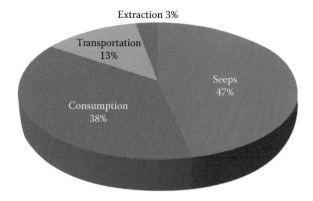

FIGURE 1.4
The breakdown of oil input into the sea.

PHOTO 1.2
The 2010 *Deep Water Horizon* was the second worst spill on record. This photo shows oil rising from the 1500-meter well blowout. The vessels in the background are working on relief well drilling and well control. (Photo from the U.S. Coast Guard Web site: http://cgvi.uscg.mil.)

PHOTO 1.3
The 1979 IXTOC blowout was the third worst spill on record. This oil well blowout in Mexican waters released water-in-oil emulsion to the surface along with lighter fuel components, which burned. (Photo from Al Allen.)

PHOTO 1.4
The *Selendang Ayu* vessel broken in two off Dutch Harbor in Alaska. This fishing vessel ran into problems with weather. The white material is sea foam while the brown material is Bunker oil. (Photo courtesy of the International Tanker Owners Pollution Federation [ITOPF].)

PHOTO 1.5
The *Amoco Cadiz* ship and spill off the coast of France in 1978. This was one of the worst tanker accidents. (Courtesy of International Maritime Organization.)

the high profile given these incidents in the media have contributed to this misconception. In fact, as stated earlier, half of the oil spilled in the seas is the runoff of oil and fuel from land-based sources rather than from accidental spills.

In conclusion, it is important to study spill incidents from the past to learn how the oil affected the environment, what cleanup techniques worked and what improvements can be made, and to identify the gaps in technology. Photographs of many of the incidents listed in Table 1.1 appear throughout this book as examples of oil behavior and cleanup techniques.

Some Oil Spill Facts

- About 20 million tons of oil and petroleum products are used worldwide each day. Despite the large number of oil spills, however, only a small percentage of oil used is actually spilled.
- Although most of the large oil spills in the marine environment are from tankers, these spills make up less than about 5% of all oil pollution entering the sea. Most oil pollution in the oceans comes from natural seeps.
- In Canada, about 12 spills of more than 4000 L are reported each day, of which only about one spill is into navigable waters.

TABLE 1.1

World's Largest Spills

No	Year	Month and Day	Ship/Incident	Country	Location	Tons (×10³)
1	1991	Jan 26	Gulf War	Kuwait	Sea Island	800
2	2010	Apr 20	Deepwater Horizon	United States	Gulf of Mexico	500
3	1979	Jun 5	IXTOC blowout	Mexico	Gulf of Mexico	470
4	1979	Jul 19	*Atlantic Empress/ Aegean Captain*	Off Tobago	Caribbean Sea	287
5	1992	Mar 2	Oil well blowout	Uzbekistan	Fergana Valley	285
6	1993	Feb 4	Oil platform blowout	Iran	Nowruz field	270
7	1983	Aug 6	*Castillo de Bellver*	South Africa	Saldanha Bay	260
8	1991	May 28	*ABT Summer*	Off Angola	Atlantic Ocean	260
9	1978	Mar 16	*Amoco Cadiz*	France	Brittany	223
10	1991	Apr 11	*Haven*	Italy	Genoa	144
11	1980	Aug 11	Oil well blowout	Libya	Inland	140
12	1988	Nov 10	*Odyssey*	Off Canada	North Atlantic	132
13	1967	Mar 18	*Torrey Canyon*	England	Lands End	119
14	1972	Dec 19	*Sea Star*	Oman	Gulf of Oman	115
15	1981	Aug 20	Storage tanks	Kuwait	Shuaybah	110
16	1971	Dec 7	*Texaco Denmark*	Belgium	North Sea	107
17	1994	Oct 25	Pipeline rupture	Russia	Usinsk	105
18	1976	May 12	*Urquiola*	Spain	La Coruna	100
19	1978	May 25	Pipeline rupture	Iran	Ahvazin	100
20	1980	Feb 23	*Irenes Serenade*	Greece	Pylos	100
21	1969	Feb 11	*Julius Schindler*	Portugal	Azores	95
22	1977	Feb 23	*Hawaiian Patriot*	Off United States	West of Hawaii	95
23	1979	Nov 15	*Independentza*	Turkey	Bosporus Strait	95
24	1975	Jan 29	*Jakob Maersk*	Portugal	Oporto	88
25	1979	Jul 6	Storage tanks	Nigeria	Forcados	85
26	1993	Jan 5	*Braer*	United Kingdom	Shetland Islands	85
27	1989	Dec 19	*Kark 5*	Morocco	Atlantic Ocean	80
28	1992	Dec 3	*Aegean Sea*	Spain	La Coruna	75
29	1985	Dec 6	*Nova*	Iran	Persian Gulf	72
30	1992	Apr 17	*Katina P*	South Africa	Indian Ocean	72
31	1996	Feb 15	*Sea Empress*	United Kingdom	Milford Haven	72
32	1971	Feb 27	*Wafra*	South Africa	Atlantic Ocean	70
33	1978	Dec 11	Fuel storage tanks	Rhodesia	Salisbury	65
34	2002	Nov 13	*Prestige*	Spain	Off Spain	63

TABLE 1.1 (*Continued*)
World's Largest Spills

No	Year	Month and Day	Ship/Incident	Country	Location	Tons (×10³)
35	1960	Dec 6	*Sinclair Petrolore*	Brazil	Off Brazil	60
36	1970	Mar 20	*Othello*	Sweden	Vaxholm	60
37	1975	May 13	*Epic Colocotronis*	United States	West of Puerto Rico	60
38	1978	Jun 12	Fuel storage tanks	Japan	Sendai	60
39	1974	Nov 9	*Yuyo Maro 10*	Japan	Tokyo	54
40	1983	Jan 7	*Assimi*	Oman	Ras al Had	53
41	1965	May 22	*Heimvard*	Japan	Hokkaido	50
42	1978	Dec 31	*Andros Patria*	Spain	Bay of Biscay	50
43	1968	Jun 13	*World Glory*	South Africa	Indian Ocean	48
44	1983	Dec 9	*Peracles GC*	Qatar	Persian Gulf	48
45	1974	Aug 9	*Metula*	Chile	Strait of Magellan	47
46	1970	Jun 1	*Ennerdale*	Seychelles	Indian Ocean	46
47	1975	Jan 13	*British Ambassador*	Japan	Iwo Jima	46
48	1994	Oct 21	*Thanassis A*	Hong Kong	South China Sea	46
49	1978	Dec 7	*Tadotsu*	Indonesia	Strait of Malacca	44
50	1968	Feb 29	*Mandoil*	United States	Oregon	43
51	1974	Dec 18	Storage tanks	Japan	Mizushima refinery	39
52	1979	Aug 26	*Patianna*	United Arab Emirates	Dubai	38
53	1972	Jun 11	*Trader*	Greece	Mediterranean Sea	37
54	1980	Dec 29	*Juan Antonio Lavalleja*	Algeria	Arzew	37
55	1988	Apr 22	*Athenian Venture*	Off Canada	Atlantic Ocean	37
56	1989	Mar 24	*Exxon Valdez*	United States	Valdez, Alaska	37
57	1973	Jun 10	*Napier*	Chile	Off west coast	36
58	1976	Feb 6	*St. Peter*	Colombia	Pacific Ocean	36
59	1978	Dec 14	Storage tanks	United States	Puerto Rico	36
60	1977	Jan 17	*Irene's Challenge*		Pacific Ocean	35
61	1978	Oct 19	Pipeline rupture	Turkey	Mardin	35
62	1979	Nov 1	*Burmah Agate*	United States	Galveston, Texas	35
63	1996	Mar 7	Unknown tanker	Mexico	Bay of Campeche	35
64	1972	Jan 28	*Golden Drake*		Northwest Atlantic	34

(*Continued*)

TABLE 1.1 (*Continued*)

World's Largest Spills

No	Year	Month and Day	Ship/Incident	Country	Location	Tons (×10³)
65	1977	Feb 7	*Borag*	Taiwan	Chilung	34
66	1986	Oct 3	Abkatun Production well blowout	Mexico	Bay of Campeche	34
67	1970	Dec 28	*Chryssi*		Northwest Atlantic	33
68	1979	Apr 28	*Gino*	France	Brittany	33
69	1968	Feb 8	*Pegasus*		North Atlantic	32
70	1969	Nov 25	*Panocean*	Taiwan	Northwest Pacific	32
71	1970	May 5	*Polycommander*	Spain	Vigo	32
72	1971	Mar	*Texaco Oklahoma*		Northwest Atlantic	32
73	1977	May 27	*Caribbean Sea*		East Pacific	32
74	1976	Jul 28	*Cretan Star*		Indian Ocean	31
75	1976	Apr 25	*Ellen Conway*	Algeria	Arzew	31
76	1969	Nov 25	*Keo*		Northwest Atlantic	30
77	1976	Dec 30	*Grand Zenith*		Northwest Atlantic	30
78	1977	Dec 16	*Venpet & Venoil*	South Africa		30
79	1979	Aug 16	*Ioannis Angelicoussis*	Angola	Malongo	30
80	1986	Apr 27	Storage tanks Texaco	Panama	Bahia las Minas	30
81	2003	July 27	*Tasman Spirit*	Pakistan	Karachi	30
82	1997	Oct 15	*Evoikos*	Singapore	Straits of Singapore	29
83	1972	Apr 1	*Guiseppe Guiljetti*		Northeast Atlantic	27
84	1977	Mar 22	Offshore Platform - Ekofisk Bravo	Norway	North Sea	27
85	1979	Jan 8	*Betelgeuse*	Ireland	Bantry Bay	27
86	1982	Nov 26	*Haralabos*	Eygpt	Ras Gharib	27
87	1976	Dec 15	*Argo Merchant*	United States	Nantucket, Massachusetts	26
88	1980	Jan 17	Funiwa #5 well blowout	Nigeria	Off Forcados	26
89	1977	Oct 28	*Al Sabbiyah*	Japan	Philippine Sea	25
90	2006	Jul 14	Jiyeh Power Station	Jordan	Off Jordan	25
91	1983	May 15	*Bellona*	Sweden	Gothenburg	24

TABLE 1.1 (*Continued*)

World's Largest Spills

No	Year	Month and Day	Ship/Incident	Country	Location	Tons (×10³)
92	1989	Dec 29	*Aragon*	Portugal	Madiera	24
93	1993	Jan 21	*Maersk Navigator*	Indonesia	Strait of Malacca	24
94	1994	Jan 24	*Cosmos A.*	Hong Kong	South China Sea	23
95	1967	Oct 1	Offshore pipeline	United States	Off Louisiana	22
96	1967	Sep	*R.C. Stoner*		North Pacific	20
97	1968	Nov 3	*Spyros Lemos*	Spain	Vigo	20
98	1972	Jun 22	Storage tanks	United States	Schuylkill River, Pennsylvania	20
99	1975	Jan 12	*Master Stathios*	South Africa	Indian Ocean	20
100	1975	Apr 4	*Spartan Lady*	United States	Atlantic Ocean	20
101	1977	Aug 10	*URSS 1*	Turkey	Bosporus Strait	20
102	1981	Mar 29	*Cavo Cambanos*	Spain	Mediterranean Sea	20
103	1993	Mar 6	Omsk-Irkutsk pipeline	Russia	West Siberia	20
104	1967	Oct	*Giorgio Fassio*	Angola	Atlantic Ocean	19
105	1976	Feb 16	*Nan Yang*	Hong Kong	South China Sea	19
106	1997	Jan 2	*Nakhodka*	Japan	Japan Sea	19
107	1985	Nov 6	Ray Richley well blowout	United States	Ranger, Texas	18
108	1991	Jul 21	*Kirki*	Australia	Cervantes	18
109	1999	Dec 12	*Erika*	France	Bay of Biscay	18
110	1970	Apr 17	*Silver Ocean*	South Africa	Durban	17
111	1980	Jan 16	*Salem*	Senegal	Atlantic Ocean	17
112	1983	Nov 26	*PNOC Basilan*	Philippines	South China Sea	17
113	1966	Feb 25	*Ann Mildred Brovig*	Germany	North Sea	16
114	1968	May 5	*Andron*	South Africa	Cape Town	16
115	1979	Mar 2	*Messiniaki Frontis*	Greece	Crete	16
116	1980	Mar 7	*Tanio*	France	Brittany	16
117	1994	Mar 31	*Seki*	United Arab Emirates	United Arab Emirates	15.9
118	1965	Aug 30	*Arsinoe*	Philippines	South China Sea	15
119	1970	Nov 1	*Marlena*	Italy	Sicily	15

(*Continued*)

TABLE 1.1 (*Continued*)

World's Largest Spills

No	Year	Month and Day	Ship/Incident	Country	Location	Tons (×10³)
120	1970	Jan 31	Gezina Brovig	United States	Puerto Rico	15
121	1976	Jun 30	Al Dammam	Greece	Agiol Theodoroi	15
122	1976	Oct 4	LSCO Petrochem	United States	Off Louisiana	15
123	1983	Nov 17	Storage tanks Shell refinery	Singapore	Puleu Bukom	15
124	1969	Jan 28	Oil well blowout	United States	Santa Barbara, California	14
125	1981	Nov 21	Globe Assimi	Lithuania	Klaipeda	14
126	1966	Oct	Malmohus	Tanzania	Dar Es Salaam	13
127	1971	Dec 2	Laban Island well blowout	Iran	Persian Gulf	13
128	1976	May 26	Storage tanks	United States	Hackensack, New Jersey	13
129	1990	Jun 8	Mega Borg	United States	Gulf of Mexico	13
130	1975	Mar 26	Tarik Ibn Ziyad	Brazil	Rio de Janeiro	12
131	1975	Jan 6	Showa Maru	Malaysia	Strait of Malacca	12
132	1980	Oct 2	Offshore platform, Hasbah 6	Saudi Arabia	Persian Gulf	12
133	1984	Oct 31	Puerto Rican	United States	San Francisco, California	12
134	1971	Jun 1	Santa Augusta	United States	Virgin Islands	11
135	1978	Feb 8	Storage tanks	United States	Los Angeles, California	11
136	1979	Jan 1	Corpoven well blowout	Venezuela	El Tigre	11
137	1983	Jan 7	Storage tanks	United States	Newark Bay, New Jersey	11
138	1987	Jun 23	Fuyoh Maru	France	Le Havre	11
139	1987	Jul 29	Blue Ridge	United States	Off Florida	11
140	1972	Aug 21	Oswego Guardian	South Africa	Indian Ocean	10
141	1974	Oct 10	Trojan	Philippines	South China Sea	10
142	1974	Sep 25	Eleftheria	Sierra Leone	Atlantic Ocean	10
143	1975	Nov 12	Olympic Alliance	United Kingdom	English Channel	10
144	1977	Nov 2	Matsushima Maru No. 3	Japan	Philippine Sea	10
145	1978	Jan 9	Brazilian Marina	Brazil	San Sebastiao	10
146	1978	Dec 25	Kosmas M.	Turkey	Asbas	10
147	1978	Jan 31	Storage tank transfer loss	United States	Arthur Kill, New Jersey	10

TABLE 1.1 (*Continued*)

World's Largest Spills

No	Year	Month and Day	Ship/Incident	Country	Location	Tons (×10³)
148	1979	Apr 5	*Fortune*	Singapore		10
149	1988	Oct 10	*Century Dawn*	Singapore	Singapore Strait	10
150	1989	Oct 4	*Pacificos*	South Africa	Indian Ocean	10
151	1990	Jun 20	Storage tanks	Russia	Nefteyugansk	10
152	1996	Nov 23	Storage tanks	Czech Republic	Litinov	10
153	1970	Feb 10	Oil well blowout, Chevron Main Pass	United States	Louisiana	9
154	1972	Sep 6	Pipeline rupture, Nipisi	Canada	Nipisi, Alberta	9
155	1978	May 6	*Eleni V*	United Kingdom	Norfolk	9
156	1984	Jul 30	*Alvenus*	United States	Cameron, Louisiana	9
157	1990	Aug 6	*Sea Spirit*	Gibraltar	Strait of Gibraltar	9
158	2007	Dec 7	*Hebei Spirit*	Korea	Off Seoul	9.4
159	1968	Mar 3	*Ocean Eagle*	United States	San Juan, Puerto Rico	8
160	1970	Feb 4	*Arrow*	Canada	Nova Scotia	8
161	1970	Dec 1	Offshore platform, Shell #26	United States	Louisiana	8
162	1972	Jan 1	*General MC Meiggs*	United States	Juan de Fuca, Washington	8
163	1977	Oct 29	*Al-Rawdatain*	Italy	Genoa	8
164	1983	Sep 27	*Sivand*	United Kingdom	Humber Estuary	8
165	1985	Apr	*Southern Cross*	Algeria	Skikda	8
166	1987	Oct 10	Yum II/ Zapoteca	Mexico	Bahia de Campeche	8
167	1990	Jun 28	*Chenki*	Egypt	Suez Canal	8
168	1992	Jun 1	Komineft Vozey pipeline	Russia	Izhma	8
169	1994	Mar 2	Oil well blowout	Uzbekistan	Fergana Valley	8
170	1972	Oct 25	*Barge Ocean 80*	United States	Arthur Kill, New Jersey	7
171	1974	Apr 7	*Sea Spirit*	United States	Los Angeles, California	7
172	1976	Oct 14	*Boehlen*	France	Atlantic Ocean	7
173	1980	Nov 22	*Georgia*	United States	Louisiana	7

(*Continued*)

TABLE 1.1 (*Continued*)

World's Largest Spills

No	Year	Month and Day	Ship/Incident	Country	Location	Tons (×10³)
174	1989	Jan 16	*UMTB American Barge 283*	United States	Off Alaska	7
175	1990	Dec 27	Kuybyshev-Perm pipeline	Russia	Cormova	7
176	1993	Oct 1	*Frontier Express*	Korea	Yellow Sea	7
177	1997	Jan 18	*Bona Fulmar*	France	Dover Strait	7
178	2000	Oct 3	*Natuna Sea*	Indonesia	Singapore Strait	7
179	1966	May 15	*Fina Norvege*	Italy	Sardinia	6
180	1973	Aug 8	Trinimar 327 well blowout	Venezuela	Güiria	6
181	1974	Feb 13	*Sea Spray*	Vietnam	South China Sea	6
182	1979	Mar 15	*Kurdistan*	Canada	Nova Scotia	6

PHOTO 1.6

A series of pipelines transports oil over the Alaskan North Slope. Pipelines are a very significant source of volume of spills on land.

PHOTO 1.7
The *Atlantic Empress* burns off the coast of Trinidad in 1979. This spill, which involved the collision between two tankers, was the fourth largest spill to date. (Photo from Al Allen.)

PHOTO 1.8
The Trans Alaska Pipeline System (TAPS) leaks oil after a puncture. (Photo from Al Allen.)

PHOTO 1.9
The *Exxon Valdez* leaking oil in 1989. This well-known incident ranks 56 on the list of 150 largest spills. (Photo from Al Allen.)

PHOTO 1.10
The *Eagle Otome* leaking oil after a collision with a barge in Port Arthur, Texas, 2010. (Photo courtesy of the International Tanker Owners Pollution Federation [ITOPF].)

In the United States, about 15 such spills occur each day into navigable waters and about 85 occur on land.

- Human error, directly or indirectly, causes 30% to 50% of oil spills; equipment failure or malfunction causes 20% to 40% of all spills.
- The average cost of cleaning up oil spills worldwide varies from $50 to $400 per liter of oil spilled.

2

Response to Oil Spills

Rapid and effective response to oil spills will result in less overall damage to the environment. Although it is important to focus on ways to prevent oil spills, methods for controlling them and cleaning them must also be rapidly and effectively implemented. An integrated system of contingency plans and response options can speed and improve response to an oil spill, and significantly reduce the environmental impact and severity of the spill.

The purpose of contingency plans is to coordinate all aspects of the response to an oil spill. This includes stopping the flow of oil, containing the oil, and cleaning it. The area covered by contingency plans could range from a single bulk oil terminal to an entire section of coastline. Oil spills, like forest fires and other environmental emergencies, are not predictable and can occur anytime and during any weather. Therefore, the key to effective response to an oil spill is to be prepared for the unexpected and to plan spill countermeasures that can be applied in the worst possible conditions.

This chapter deals with planning a response to an oil spill and the functions of oil spill response organizations in industry and government. Topics covered include contingency planning for oil spills, which encompasses the activation of such plans; the structure of response organizations, training, and supporting studies and sensitivity mapping; communications systems; oil spill cooperatives; the role of private and government response organizations; and cost recovery.

Oil Spill Contingency Plans

Studies of several major oil spills in the early 1970s showed that response to these spills suffered not only from a lack of equipment and specialized techniques, but also from a lack of organization and expertise to deal with such emergencies. Since then, contingency plans have evolved and today often cover wide areas, and pool national and even international resources and expertise.

It is now recognized that oil spills vary in size and impact and require different levels of response. Contingency plans can be developed for a particular facility, such as a bulk storage terminal, which would include organizations and resources from the immediate area, with escalating plans for spills of greater impact. Contingency plans for provinces, states, or even

PHOTO 2.1
Exercises are a very important part of preparedness. In this photo, the staff are exercising a major contingency plan. (Photo from Prince William Sound Regional Citizens Advisory Committee.)

the entire country usually focus more on roles and responsibilities and providing the basis for cooperation between the appropriate response organizations rather than on specific response actions. Some elements that may be included in contingency plans today are listed in Table 2.1. Most contingency plans usually include:

- A list of persons and agencies to be notified immediately after a spill occurs
- An organization chart of response personnel and a list of their responsibilities as well as a list of actions to be taken by them in the first few hours after the incident
- Area-specific action plans
- A communications network to ensure response efforts are coordinated among the response team
- Protection priorities for the affected areas
- Operational procedures for controlling and cleaning the spill
- Reference material such as sensitivity maps and other technical data about the area
- Procedures for informing the public and keeping records
- An inventory or database of the type and location of available equipment, supplies, and other resources
- Scenarios for typical spills and decision trees for certain types of response actions such as using chemical treating agents or in-situ burning

TABLE 2.1

Contents of Typical Contingency Plans

First Response

First actions
Initial contact
First procedures
Addresses and phone numbers

Activation

Activation procedures

Levels of Response

Actions for various levels
Escalation procedures

Organization

Responsibilities
Individuals and roles

Reporting

Systems
Procedures

Protection Priorities

Critical areas
Sensitive areas

Operations

Individuals and roles
Surveillance, monitoring, and reconnaissance
Equipment deployment
Communications
Record keeping
Public relations
Shoreline surveillance

Action Plans

Deployment areas
Shoreline assessment and countermeasures
Disposal options

Scenarios

Decision trees
Scenarios

Exercises

Exercise procedures

(Continued)

TABLE 2.1 (*Continued*)

Contents of Typical Contingency Plans

Databases

	Maps
	Contacts
	Equipment
	Lists
	Vendors
	Supplies
	Sensitivity data
	Extra resources

Science/Technology

	Modeling
	Resources
	Duties

PHOTO 2.2
Testing a new piece of equipment. This builds familiarity with the equipment and thus improves preparedness. (Photo from the Canadian Coast Guard.)

To remain effective, response options detailed in contingency plans must be frequently tested. This testing is conducted by responding to a practice spill as though it is real. This varies from a tabletop exercise to large-scale field exercises in which equipment is deployed and oil is actually "spilled" and recovered. Such exercises not only maintain and increase the skills of the response personnel but also lead to improvements and fine-tuning of the plan as weaknesses and gaps are identified.

PHOTO 2.3
Staging an oil spill boom and other equipment in preparation for a response. (Photo from the U.S. Coast Guard Web site: http://cgvi.uscg.mil.)

PHOTO 2.4
Equipment at a cooperative warehouse. The equipment is not only stored, but maintained and tested on a regular schedule.

Activation of Contingency Plans

The response actions defined in contingency plans, whether for spills at a single industrial facility or in an entire region, are separated into the following phases: alerting and reporting; evaluation and mobilization; containment and recovery; decontamination of equipment; disposal; and remediation or restoration. In practice, these phases often overlap rather than follow one another consecutively.

Most contingency plans also allow for a "tiered response," which means that response steps and plans escalate as the incident becomes more serious. As the seriousness of an incident is often not known in the initial phases, one of the first priorities is to determine the magnitude of the spill and its potential impact.

Alerting the first-response personnel and the responsible government agency is the first step in activating an oil spill contingency plan. Reporting a spill to the designated agency, regardless of the size or seriousness of the spill, is a legal requirement in most jurisdictions in Canada, the United States, and in other countries.

The first-response personnel assess the situation and initiate actions to control, contain, or minimize the environmental damage as soon as possible. Until the full command structure is in place and operating, employees carry out their responsibilities according to the contingency plan and their training. This emphasizes the need for a detailed contingency plan for this phase of the operation and the importance of a high level of training in first response.

Stopping the flow of oil is a priority in the first phase of the operation, although response may need to be immediate and be undertaken in parallel with stopping the flow. In the case of a marine accident such as a ship grounding, stopping the flow of oil may not be possible; however, the spillage may be minimized by pumping oil in the ruptured tanks into other tanks or by pumping oil from leaking tankers into other tankers or barges. These operations may take up to a week to complete and are often delayed by bad weather. Once the flow of oil has been stopped, emphasis switches to containing the oil or diverting it from sensitive areas. Typically, containment and recovery efforts parallel the flow control procedure.

As oil spills pose many dangers, safety is a major concern during the early phases of the response action. First, the physical conditions at the site may not be well known. Second, many petroleum products are flammable or contain volatile and flammable compounds, creating a serious explosion and fire hazard in the early phases of the spill. Third, spills may occur during bad weather or darkness, which increases the danger to personnel.

As more of the individuals called appear on the scene and begin to take up their duties, the response plan falls into place. Response strategies vary from incident to incident and in different circumstances and take varying amounts of time to carry out. Response to a small spill may be fully operational within hours, whereas for a larger spill, response elements such as shoreline assessment after cleanup may not take place until weeks after the incident.

Training

A high-caliber training program is vital for a good oil spill response program. Response personnel at all levels require training in specific operations and on using equipment for containing and cleaning up spills. To minimize injury during response, general safety training is also crucial. In many countries, response personnel are required to have 12 to 40 hours of safety training before they can perform fieldwork.

Ongoing training and refresher courses are also essential in order to maintain and upgrade skills. Training techniques for spill response include tools such as audiovisuals and computer simulation programs, which make the training more realistic and effective.

Structure of Response Organizations

Most contingency plans define the structure of the response organization so that roles and command sequences are fully understood before any incident occurs. The on-scene commander (OSC) is the head of the response effort and is experienced in oil spill response operations. The OSC is responsible for making all major decisions on actions taken. This person ensures that the various aspects of the operation are coordinated and sequenced and that a good communications system is in place.

The OSC is supported by a fully trained staff or response team whose duties are clearly defined in the contingency plans. One or more individuals are often designated as deputy on-scene commanders to ensure that there is backup for the OCS and that multiple shifts can be run.

The structure of a typical spill response organization is shown in Figure 2.1. A command structure specified in the United States is a system called the Incident Command System, or ICS. This basic structure is shown in Figure 2.2, and involves common elements to ensure uniformity across organizations and to make it easier for federal responders to deal with

PHOTO 2.5
A typical command post. Members are listening to a daily briefing. (Photo from Ed Stanton.)

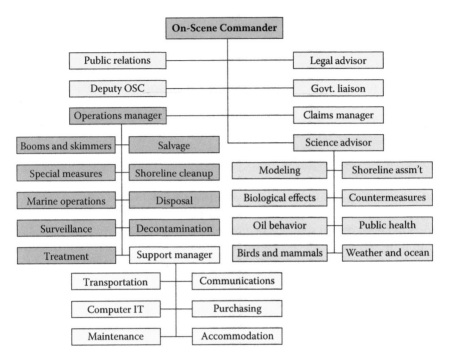

FIGURE 2.1
A typical response organization chart.

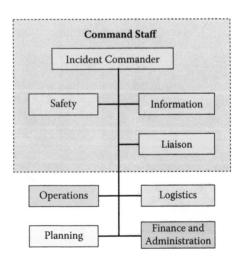

FIGURE 2.2
The Incident Command System (ICS) basic structure.

contingency plans in areas other than their own familiar territories. The Unified Command System, or UCS, is similar to ICS, but involves the joining of the company, state or province, and federal response structures. The idea is to join forces to maximize the resources available to deal with the spill and to avoid duplication.

The success of any response operation, contingency plan, and organizational structure depends primarily on the level of commitment of both the response personnel and the response organization itself. The training, experience, and capabilities of the response personnel in their respective functions and their ability to work as a team are also crucial to the success of the response operation.

The care and effort taken in developing the plan is also important to its success. In addition, the response team and the plan itself must be flexible enough to accommodate different sizes of spills and different circumstances. And finally, sufficient resources must be available to prepare and implement the plan, and to carry out frequent testing of the plan.

Supporting Studies and Sensitivity Mapping

A contingency plan usually includes background information on the area covered by the plan. This consists of data collected from studies and surveys and often takes the form of a sensitivity map for the area.

As shown in Table 2.2, sensitivity maps contain information on potentially sensitive physical and biological resources that could be affected by an oil spill. This includes concentrations of wildlife such as mammals, birds, and fish; human amenities, such as recreational beaches; natural features such as water currents and sandbars; and types of shorelines. Features that are important for spill response, such as roads and boat launches, are also included.

Sensitivity maps are now computerized in systems called geographic information systems (GIS). These systems allow a composite map or image to be drawn in layers. Bird populations, for example, typically constitute one layer of a GIS map. This allows personnel to rapidly update and analyze data in the area. Detailed information is usually kept in tables as part of the GIS. Sensitivity maps can also be integrated with computerized oil spill models so that the impact of an oil spill on the environment can be projected.

Other types of studies that may be included in the sensitivity map are area-specific response strategies, such as studies on using booms as a containment strategy in a certain area; information on tides, currents, and water circulation patterns; climatic studies to yield prevailing winds for a given area; detailed shoreline surveys and data sets; spill probabilities and statistics for a region; disposal sites for oily debris; inventories of countermeasures equipment and cleanup personnel; and detailed logistic surveys or plans to assist in transporting equipment.

TABLE 2.2

Elements of Sensitivity Maps

Most Sensitive Areas

Priority protection areas

Most sensitive habitats

Biota

Seasonal locations of:

Birds

Mammals

Fish

Sensitive plants

Important natural resources

Environmental Features

Wetlands

Reserves

Unusual features

Water currents

Shoreline

Types

Sensitivities

Vegetation types

Spill Control Features

Boom placement locations

Equipment locations

Zones delineated for countermeasures

Priority cleanup areas

Human Amenities Requiring Protection

Water intakes

Waterfront properties

Recreational beaches

Parks and reserves

Fish traps and ponds

Mariculture areas

Marinas

Archaeological sites

Physical Assets

Roads

Sewer outfalls

Boat launches

Dams and weirs

Communication Systems

Good communications are essential in an oil spill response operation. Field personnel must be in constant contact for both operational and safety reasons. Field personnel use both VHF and UHF radios for spoken communication. Several frequencies have been assigned exclusively for spill response crews. The range of these radios varies with the terrain but is generally less than 30 km. Of course, cellular telephones and smart phones now dominate the communications methods. Communications on cell phones may not be secure and is less reliable than dedicated radio systems.

Some response organizations have established relay stations to ensure coverage over their entire area. Direct satellite communications is also frequently used in spill operations, but due to the size of equipment required, is usually limited to base stations. Arrangements are often made with telephone companies to provide dozens of lines into the command posts when a major spill occurs. Fax machines are still an important means of communications in response operations. This is replaced in modern times by sending attachments along with e-mails. Many response trucks and boats are equipped with fax receivers and satellite communications.

Arrangements are often made with other response organizations to move equipment into the spill area. For rapid deployment of this equipment when a spill occurs, databases of information about the equipment and the logistics of moving it must be referred to in advance. Arrangements are already in place with many airfreight and trucking companies to move equipment when a major spill occurs.

PHOTO 2.6
A mobile command post. This enables the coordination of a response directly at the scene as well as coordination with the central command post. (Photo from the U.S. Coast Guard Web site: http://cgvi.uscg.mil.)

Oil Spill Cooperatives

As most oil companies or firms that handle oil do not have staff dedicated to cleaning oil spills, several companies in the same area often join forces to form cooperatives. By pooling resources and expertise, these oil spill cooperatives can then develop effective and financially viable response programs. The cooperative purchases and maintains containment, cleanup, and disposal equipment, and provides the training for its use.

A core of trained people is available for spill response and other response personnel can quickly be hired on a casual basis for a large spill. Neighboring cooperatives also join forces to share equipment, personnel, and expertise. Oil spill cooperatives vary in size but are usually made up of about 10 full-time employees and a million dollars worth of equipment, and cover an area of several thousand square kilometers.

In recent years, very large cooperative response organizations have been formed that cover entire countries. In Canada, the Eastern Canada Response Corporation (ECRC) has developed response depots across marine waters and through the St. Lawrence Seaway. Burrard Clean has similarly set up cleanup depots in British Columbia and another cooperative has been organized for the Prairie provinces. In the United States, the Marine Spill Response Corporation (MSRC) and the National Response Corporation (NRC) have similar capabilities. These organizations involve as many as 300 full-time employees and over a hundred million dollars worth of equipment. Some large international response organizations, such as Oil Spill Response and East Asia Response Limited (OSRL/EARL), have also been formed in Southampton, England, and in Singapore, Malaysia.

Private and Government Response Organizations

In North America, private firms also provide oil spill containment and cleanup services. These firms are often also engaged in activities such as towing, marine salvage, or waste oil disposal, and sometimes operate remote cleanup operations or maintain equipment depots as well. Many of these firms have contracts with the cooperatives to provide services. Private firms can often recruit large numbers of cleanup personnel on short notice and are valuable allies to industry and government organizations dealing with spills. Their resources are often included in local and regional contingency plans.

Government response organizations, such as the Coast Guard, often have large stockpiles of equipment and trained personnel. They often respond to a spill when there are no responsible parties or before full response capabilities have been organized. The Coast Guard in Canada and the United States also provide rapid response for lightering (unloading) stricken tankers and dealing with sunken vessels, which the private sector generally cannot do.

Government organizations are often responsible for monitoring cleanup operations to ensure that measures taken are adequate and that environmental damage is minimal. Environment Canada has set up Regional Environmental Emergencies Teams (REETs) and the U.S. Environmental Protection Agency (EPA) has established Regional Response Teams (RRTs) to coordinate the environmental aspects of spill response. These teams are made up of members from various federal and provincial/state organizations. Government agencies have significant resources that can be incorporated into response efforts. These include scientific expertise, on-site and laboratory services, and monitoring instruments to measure parameters related to health and safety issues.

In some spill situations, especially large spills, volunteers are an important part of the response effort. Volunteers are usually trained and given accommodations, and their efforts coordinated with the main spill cleanup.

PHOTO 2.7

A training course in shoreline assessment. The class is using a real spill for this course. (Photo from the U.S. Coast Guard Web site: http://cgvi.uscg.mil.)

PHOTO 2.8
Volunteers clean up a shoreline in Korea. This spill cleanup was carried out by hundreds of volunteers. (Photo courtesy of the International Tanker Owners Pollution Federation [ITOPF].)

Cost Recovery

Many response organizations now function under cost-recovery schemes, charging back the costs of their services to the companies that actually own them, to clients who use their services, or to neighboring cooperatives or nonmembers. Cost recovery has actually made it easier for cooperatives to work outside their areas of responsibility.

All petroleum shipping agencies are covered by some form of insurance, which allows spill operations to be conducted under cost recovery. Insurance firms, sometimes also called indemnity clubs, have experts who monitor spill control operations to ensure that cleanup is conducted cost effectively. International agreements are in place to ensure that tanker owners are covered by insurance and that certain minimum standards for oil spill cleanup are maintained.

3

Types of Oil and Their Properties

Oil is a general term that describes a wide variety of natural substances of plant, animal, or mineral origin, as well as a range of synthetic compounds. The many different types of oil are made up of hundreds of major compounds and thousands of minor ones. This chapter discusses petroleum oils and petroleum products. As their composition varies, each type of oil or petroleum product has certain unique characteristics or properties. These properties influence how the oil behaves when it is spilled and determine the effects of the oil on living organisms in the environment. These properties also influence the efficiency of cleanup operations. This book deals specifically with crude oils and petroleum products derived from crude oils. The chemical composition and physical properties of these oils are described in this chapter.

The Composition of Oil

Crude oils are mixtures of hydrocarbon compounds ranging from smaller, volatile compounds to very large, nonvolatile compounds. This mixture of compounds varies according to the geological formation of the area in which the oil is found and strongly influences the properties of the oil. For example, crude oils that consist primarily of large compounds are viscous and dense. Petroleum products such as gasoline or diesel fuel are mixtures of fewer compounds and thus their properties are more specific and less variable.

Hydrocarbon compounds are composed of hydrogen and carbon, which are the main elements in oils. Oils also contain varying amounts of sulfur, nitrogen, oxygen, and sometimes mineral salts, as well as trace metals such as nickel, vanadium, and chromium.

In general, the hydrocarbons found in oils are characterized by their structure. The hydrocarbon structures found in oil are the saturates, aromatics, and polar compounds, some examples of which are shown in Figure 3.1.

The **saturate group** of components in oils consists primarily of alkanes, which are compounds of hydrogen and carbon with the maximum number of hydrogen atoms around each carbon. Thus, the term *saturate* is used because the carbons are "saturated" with hydrogen. The saturate group also includes cycloalkanes, which are compounds made up of the same carbon

Groupings	Example Classes, Names, and Compounds			
	Chemical Class	Alternate Name	Description	Example Compound
Saturates	Alkanes	Paraffins		Dodecane  $C_{12}H_{26}$
	Cycloalkanes Waxes	Naphthenes	N-alkanes C_{18}-C_{80}	Decalin
Aromatics	BTEX		Benzene, toluene, ethylbenzene, xylenes	Benzene
	PAHs			Anthracene 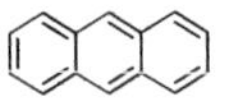
	Naphthenoaromatics		Combinations of aromatics and cycloalkanes	Tetralin 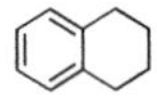
Resins	Class of mostly anomalous polar compounds sometimes containing oxygen, nitrogen, sulfur, or metals			Carbazole 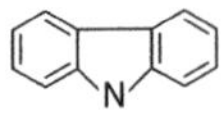
Asphaltenes	Class of large anomalous compounds sometimes containing oxygen, nitrogen, metals, or sulfur			Structures not known

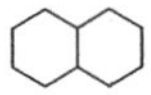

FIGURE 3.1
Chemical compounds in oil.

and hydrogen constituents but with the carbon atoms bonded to each other in rings or circles. Larger saturate compounds are often referred to as **waxes**.

The **olefins**, or unsaturated compounds, are another group of compounds that contain less hydrogen atoms than the maximum possible. Olefins have at least one double carbon-to-carbon bond, which displaces two hydrogen atoms. Significant amounts of olefins are found only in refined products.

The **aromatics** include at least one benzene ring of six carbons. Three double carbon-to-carbon bonds float around the ring and add stability. Because of this stability, benzene rings are very persistent and can have toxic effects on the environment.

The most common smaller and more volatile compounds found in oil are often referred to as *BTEX*, or benzene, toluene, ethylbenzene, and xylenes. **Polyaromatic hydrocarbons**, or **PAHs**, are compounds consisting of at least two benzene rings. PAHs make up between 0% and 60% of the composition of oil.

Polar compounds are those that have a significant molecular charge as a result of bonding with compounds such as sulfur, nitrogen, or oxygen. The polarity or charge that the molecule carries results in behavior that is different than that of unpolarized compounds, under some circumstances. In the petroleum industry, the smallest polar compounds are called **resins**, which are largely responsible for oil adhesion. The larger polar compounds are called **asphaltenes** and they often make up the largest percentage of the asphalt commonly used for road construction. Asphaltenes often have very large molecules and, if in abundance in an oil, they have a significant effect on oil behavior. This will be discussed in Chapter 4.

The following are the oils used in this book to illustrate the fate, behavior, and cleanup of oil spills:

1. Gasoline, as used in automobiles
2. Diesel fuel, as used in trucks, trains, and buses
3. Light crude oil, as produced in great abundance in many countries
4. Heavy crude oil, as produced in many countries; specific examples will be given
5. Intermediate fuel oil (IFO), a mixture of a heavy residual oil and diesel fuel used primarily as a propulsion fuel for ships (the intermediate refers to the fact that the fuel is between a diesel fuel and a heavy residual fuel)
6. Bunker fuel, such as Bunker C, which is a heavy residual fuel remaining after the production of gasoline and diesel fuel in refineries and often used in heating plants
7. Crude oil emulsion, such as an emulsion of water in a medium crude oil

Typical amounts of hydrocarbon compounds found in these oils are shown in Table 3.1.

Properties of Oil

The properties of oil discussed here are viscosity, density, specific gravity, solubility, flash point, pour point, distillation fractions, interfacial tension, and vapor pressure. These properties for the oils discussed in this book are listed in Table 3.2.

Viscosity is the resistance to flow in a liquid. The lower the viscosity, the more readily the liquid flows. For example, water has a low viscosity and flows readily, whereas honey, with a high viscosity, flows poorly. The viscosity of the oil is largely determined by the amount of lighter and heavier fractions that it contains. The greater the percentage of light components such as saturates and the lesser the amount of asphaltenes, the lower the viscosity.

As with other physical properties, viscosity is affected by temperature, with a lower temperature giving a higher viscosity. For most oils, the viscosity varies as the logarithm of the temperature, which is a very significant variation. Oils that flow readily at high temperatures can become a slow-moving, viscous mass at low temperatures. In terms of oil spill cleanup,

TABLE 3.1

Composition of Some Oils and Petroleum Products

Group	Compound Class	Gasoline	Diesel	Light Crude	Heavy Crude	IFO[a]	Bunker C
Saturates		50 to 60	65 to 95	55 to 90	25 to 80	25 to 35	20 to 30
	Alkanes	45 to 55	35 to 45	40 to 85	20 to 60	10 to 25	10 to 20
	Cycloalkanes	5	25 to 50	5 to 35	0 to 10	0 to 5	0 to 5
Olefins		5 to 10	0 to 10				
Aromatics		25 to 40	5 to 25	10 to 35	15 to 40	40 to 60	30 to 50
	BTEX	15 to 25	0.5 to 2.0	0.1 to 2.5	0.01 to 2.0	0.05 to 1.0	0.00 to 1.0
	PAHs		0 to 5	10 to 35	15 to 40	30 to 50	30 to 50
Polar Compounds				1 to 15	5 to 40	15 to 25	10 to 30
	Resins		0 to 2	0 to 10	2 to 25	10 to 15	10 to 20
	Asphaltenes		0 to 2	0 to 10	0 to 20	5 to 10	5 to 20
Sulfur		0.02	0.1 to 0.5	0 to 2	0 to 5	0.5 to 2.0	2 to 4
Metals				30 to 250	100 to 500	100 to 1000	100 to 2000

Note: Numbers in percent (%), except for metals, which is in parts per million.

[a] IFO is intermediate fuel oil, which is the residual fraction diluted by diesel fractions.

PHOTO 3.1
A highly weathered Bunker C oil that has been exposed for more than 25 years. The interior of the oil is not as weathered and resembles the original oil spilled.

PHOTO 3.2
Bits of water-in-oil emulsion floating on the water. This photo was taken close to a boat.

viscosity can affect the oil's behavior. Viscous oils do not spread rapidly, do not penetrate soil as readily, and affect the ability of pumps and skimmers to handle the oil.

Density is the mass (weight) of a given volume of oil and is typically expressed in grams per cubic centimeter (g/cm^3). It is the property used by the petroleum industry to define light or heavy crude oils. Density is also important as it indicates whether a particular oil will float or sink in water.

TABLE 3.2

Typical Oil Properties

Property	Units	Gasoline	Diesel	Light Crude	Heavy Crude	Intermediate Fuel Oil	Bunker C
Viscosity	mPa·s at 15°C	0.5	2	5 to 50	50 to 50,000	1000 to 15,000	10,000 to 50,000
Density	g/mL at 15°C	0.72	0.84	0.78 to 0.88	0.88 to 1.00	0.94 to 0.99	0.96 to 1.04
Flash point	°C	−35	45	−30 to 30	−30 to 60	80 to 100	>100
Solubility in water	ppm	200	40	10 to 50	5 to 30	10 to 30	1 to 5
Pour point	°C	NR	−35 to −10	−40 to 30	−40 to 30	−10 to 10	5 to 20
API gravity		65	35	30 to 50	10 to 30	10 to 20	5 to 15
Interfacial tension	mN/m at 15°C	27	27	10 to 30	15 to 30	25 to 30	25 to 35
Distillation fractions	% distilled at						
	100°C	70	1	2 to 15	1 to 10		
	200°C	100	30	15 to 40	2 to 25	2 to 5	2 to 5
	300°C		85	30 to 60	15 to 45	15 to 25	5 to 15
	400°C		100	45 to 85	25 to 75	30 to 40	15 to 25
	Residual			15 to 55	25 to 75	60 to 70	75 to 85

Note: NR, not relevant.

PHOTO 3.3
An environmental officer samples a fuel oil spill that occurred under the snow. The yellow snow is fuel mixed with snow.

As the density of water is 1.0 g/cm^3 at 15°C and the density of most oils ranges from 0.7 to 0.99 g/cm^3, most oils will float on water. As the density of seawater is 1.03 g/cm^3, even heavier oils will usually float on it. The density of oil increases with time, as the light fractions evaporate.

Occasionally, when the density of an oil becomes greater than the density of freshwater or seawater, the oil will sink. Sinking is rare, however, and happens only with a few oils, usually residual oils such as Bunker C. Significant amounts of oil have sunk in only about 25 incidents out of thousands.

Another measure of density is **specific gravity**, which is an oil's relative density compared to that of water at 15°C. It is the same value as density at the same temperature. Another gravity scale is that of the American Petroleum Institute (API). The **API gravity** is based on the density of pure water, which has an arbitrarily assigned API gravity value of 10°. Oils with progressively lower specific gravities have higher API gravities.

The following is the formula for calculating API gravity:

$$\text{API gravity} = (141.5 \div \text{density at } 15.5°C) - 131.5$$

Oils with high densities have low API gravities and vice versa. In the United States, the price of a specific oil may be based on its API gravity, as well as other properties of the oil.

Solubility in water is the measure of how much of an oil will dissolve in the water column on a molecular basis. Solubility is important in that the soluble fractions of the oil are sometimes toxic to aquatic life, especially at higher concentrations. As the amount of oil lost to solubility is always small, this is not as great a loss mechanism as evaporation. In fact, the solubility of oil in water is so low (generally less than 100 parts per million) that it would be the equivalent of approximately one grain of sugar dissolving in a cup of water.

The **flash point** of an oil is the temperature at which the liquid gives off sufficient vapors to ignite upon exposure to an open flame. A liquid is considered to be flammable if its flash point is less than 60°C. There is a broad range of flash points for oils and petroleum products, many of which are considered flammable, especially when fresh. Gasoline, which is flammable under all ambient conditions, poses a serious hazard when spilled. Many fresh crude oils have an abundance of volatile components and may be flammable for as long as one day until the more volatile components have evaporated. On the other hand, Bunker C and heavy crude oils typically are not flammable when spilled.

The **pour point** of an oil is the temperature at which it takes longer than a specified time to pour from a standard measuring vessel. As oils are made up of hundreds of compounds, some of which may still be liquid at the pour point, the pour point is not the temperature at which the oil will no longer pour. The pour point represents a consistent temperature at which an oil will pour very slowly and therefore has limited use as an indicator of the state of the oil. In fact, pour point has been used too much in the past to predict how oils will behave in the environment. For example, waxy oils can have very low pour points, but may continue to spread slowly at that temperature and can evaporate to a significant degree.

Distillation fractions of an oil represent the fraction (generally measured by volume) of an oil that is boiled off at a given temperature. This data is obtained on most crude oils so that oil companies can adjust parameters in their refineries to handle the oil. This data also provides environmentalists with useful insights into the chemical composition of oils. For example, whereas 70% of gasoline will boil off at 100°C, only about 5% of a crude oil will boil off at that temperature and an even smaller amount of a typical Bunker C. The distillation fractions correlate strongly to the composition as well as to other physical properties of the oil.

The oil–water **interfacial tension**, sometimes called surface tension, is the force of attraction or repulsion between the surface molecules of oil and water. Together with viscosity, surface tension is an indication of how rapidly and to what extent an oil will spread on water. The lower the interfacial tension with water, the greater the extent of spreading. In actual practice, the interfacial tension must be considered along with the viscosity because it has been found that interfacial tension alone does not account for spreading behavior.

The **vapor pressure** of an oil is a measure of how the oil partitions between the liquid and gas phases, or how much vapor is in the space above a given amount of liquid oil at a fixed temperature. Because oils are a mixture of many compounds, the vapor pressure changes as the oil weathers. Vapor pressure is difficult to measure and is not frequently used to assess oil spills.

PHOTO 3.4
A Bunker C spill from a ship fouls a saltwater marsh.

PHOTO 3.5
Heavy Bunker oil fouls a shoreline. This weathered oil is now so solid it must be cut and removed a piece at a time.

Correlation between Properties

Although there is a high degree of correlation between the various properties of an oil, these correlations should be used cautiously as oils vary so much in composition. For example, the density of many oils can be predicted based on their viscosity. For other oils, however, this could result in errors. For example, waxy oils have much higher viscosities than would be implied from their densities. There are several mathematical equations for predicting one property of an oil from another property, but these must be used carefully as there are many exceptions.

4

Behavior of Oil in the Environment

When oil is spilled, whether on water or land, a number of transformation processes occur, which are referred to as the "behavior" of the oil. Two types of transformation processes are discussed in this chapter. The first is weathering, a series of processes whereby the physical and chemical properties of the oil change after the spill. The second is a group of processes related to the movement of oil in the environment. Spill modeling is also included in the section on oil movement. Weathering and movement processes can overlap, with weathering strongly influencing how oil is moved in the environment. These processes depend very much on the type of oil spilled and the weather conditions during and after the spill.

The Importance of Behavior and Fate

The specific behavior processes that occur after an oil spill determine how the oil should be cleaned up and its effect on the environment. For example, if an oil evaporates rapidly, cleanup is less intense, but the hydrocarbons in the oil enter the atmosphere and cause air pollution. An oil slick could be carried by surface currents or winds to a bird colony or to a shore where seals or sea lions are breeding, and severely affect the wildlife and their habitat. On the other hand, a slick could be carried out to sea where it disperses naturally and has little direct effect on the environment.

In fact, the fate and effects of a particular spill are determined by the behavior processes, which are in turn almost entirely determined by the type of oil and the environmental conditions at the time of the spill. Spill responders need to know the ultimate fate of the oil in order to take measures to minimize the overall impact of the spill.

An Overview of Weathering

Oil spilled on water undergoes a series of changes in physical and chemical properties that in combination are termed **weathering**. Weathering processes occur at very different rates, but begin immediately after oil is spilled into

PHOTO 4.1
Highly evaporated oil on a beach forming an *asphalt pavement* along with gravel and oyster shells. (Photo courtesy of the Canadian Coast Guard.)

the environment. Weathering rates are not consistent throughout the duration of an oil spill and are usually highest immediately after the spill.

Both weathering processes and the rates at which they occur depend more on the type of oil than on environmental conditions. Most weathering processes are highly temperature dependent, however, and will often slow to insignificant rates as temperatures approach zero degrees.

The processes included in weathering are evaporation, emulsification, natural dispersion, dissolution, photooxidation, sedimentation, adhesion to materials, interaction with mineral fines, biodegradation, and the formation of tar balls. These processes are listed in order of importance in terms of their effect on the percentage of total mass balance, that is, the greatest loss from the slick in terms of percentage, and what is known about the process.

Evaporation

Evaporation is usually the most important weathering process. It has the greatest effect on the amount of oil remaining on water or land after a spill. Over a period of several days, a light fuel such as gasoline evaporates completely at temperatures above freezing, whereas only a small percentage of a heavier Bunker C oil evaporates. The evaporation rates of the oils discussed in this book are shown in Figure 4.1.

The rate at which an oil evaporates depends primarily on the oil's composition. The more volatile components an oil or fuel contains, the greater the extent and rate of its evaporation. Many components of heavier oils will not evaporate at all, even over long periods of time and at high temperatures.

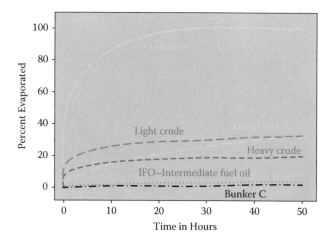

FIGURE 4.1
Evaporation of typical oils at 15°C.

Oil and petroleum products evaporate in a slightly different manner than water, and the process is much less dependent on wind speed and surface area. Oil evaporation can be considerably slowed, however, by the formation of a crust or skin on top of the oil. This happens primarily on land where the oil layer does not mix with water. The skin or crust is formed when the smaller compounds in the oil are removed leaving some compounds, such as waxes and resins, at the surface. These then seal off the remainder of the oil and prevent evaporation. Stranded oil from old spills has been reexamined over many years and it has been found that when this crust has formed, there is no significant evaporation in the oil underneath. When this crust has not formed, the same oil could be weathered to the hardness of wood.

The rate of evaporation is very rapid immediately after a spill and then slows considerably. About 80% of evaporation occurs in the first 2 days after a spill, which can be seen in Figure 4.1. The evaporation of most oils follows a logarithmic curve with time. Some oils such as diesel fuel, however, evaporate as the square root of time, at least for the first few days. This means that the evaporation rate slows very rapidly with time in both cases.

The properties of an oil can change significantly with the extent of evaporation. If about 40% of an oil evaporates, its viscosity could increase by as much as a thousandfold. Its density could rise by as much as 10% and its flash point by as much as 400%. The extent of evaporation can be the most important factor in determining properties of an oil at a given time after the spill and in changing the behavior of the oil.

Emulsification and Water Uptake

Water can enter oil through several processes. **Emulsification** is the process by which one liquid is dispersed into another one in the form of small

PHOTO 4.2
Emulsion in a boom being recovered by a weir skimmer. (Photo from the U.S. Coast Guard
Web site: http://cgvi.uscg.mil.)

PHOTO 4.3
Oil undergoing natural dispersion. This spill was of a very light crude oil.

droplets. To be called an emulsion, the product must have a certain amount
of stability. Otherwise the process is called water uptake, not emulsification.
Water droplets can remain in an oil layer in a stable form and the resulting material is completely different. These stable water-in-oil emulsions are
sometimes called mousse or **chocolate mousse** as they resemble this des-
sert. In fact, both the tastier version of chocolate mousse and butter are common examples of water-in-oil emulsions. Table 4.1 lists the five ways that oil
uptakes water and the resulting properties after water is in the oil.

TABLE 4.1

Five Ways Oils Uptake Water

Type	Mechanism	Starting Oil Characteristics	Requirements	After Water Uptake			Typical Viscosity Increase[a]	Typical Water Uptake
				Color	Typical Lifetime			
Soluble oil	Solubility	Most		Same	Years		1	<1%
Unstable or does not uptake water	None	Many oils		Same			1	—
Mesostable	Viscous entrainment and A/R interaction	Moderate viscosity and some A/R	Sea energy	Reddish until broken	3 to 6 days		50	50% to 70%
Stable	Viscous entrainment and A/R interaction	Moderate viscosity and some A/R	Sea energy	Reddish	months		800 to 1000	60% to 80%
Entrained	Viscous entrainment		Sea energy	As oil	2 to 10 days		2 to 5	30% to 40%

Note: A/R, asphaltenes and resins.

[a] Viscosity increase from starting oil.

The understanding of emulsion formation is just in the early phases, but it probably starts with sea energy forcing the entry of small water droplets, about 10 to 25 µm (or 0.010 to 0.025 mm) in size, into the oil. If the oil is only slightly viscous, these small droplets will not leave the oil quickly. On the other hand, if the oil is too viscous, droplets will not enter the oil to any significant extent. Once in the oil, the droplets slowly gravitate to the bottom of the oil layer. Any asphaltenes and resins in the oil will interact with the water droplets to stabilize them. Depending on the quantity of asphaltenes and resins, an emulsion may be formed. The conditions required for emulsions of any stability to form may only be reached after a period of evaporation. Evaporation lowers the amount of low-molecular weight compounds and increases the viscosity to the critical value.

Water can be present in oil in five ways. First, some oils contain about 1% water as soluble water. This water does not significantly change the physical or chemical properties of the oil. The second way is when water droplets are not held in the oil long enough to form an emulsion. These are called oils that do not form any type of water-in-oil mixtures or *unstable emulsions*. These are formed when water droplets are incorporated into oil by the sea's wave action and there are not enough asphaltenes and resins in the oil or if there is an insufficient viscosity to prevent droplets from leaving the oil mass. Unstable emulsions break down into water and oil within minutes or a few hours at most, once the sea energy diminishes. The properties and appearance of the unstable emulsion are almost the same as those of the starting oil, although the water droplets may be large enough to be seen with the naked eye. Unstable emulsions are also known as oils that do not uptake water to any extent.

Mesostable emulsions represent the third way water can be present in oil. These are formed when the small droplets of water are stabilized to a certain extent by a combination of the viscosity of the oil and the interfacial action of asphaltenes and resins. For this to happen, the asphaltene or resin content of the oil must be at least 3% by weight. The viscosity of mesostable emulsions is 20 to 80 times higher than that of the starting oil. These emulsions generally break down into oil and water or sometimes into water, oil, and emulsion remnants within a few days. Mesostable emulsions are viscous liquids that are reddish-brown in color, until broken.

The fourth way that water exists in oil is in the form of *stable emulsions*. These form in a way similar to mesostable emulsions except that the oil contains sufficient asphaltenes and resins. The viscosity of stable emulsions is 800 to 1000 times higher than that of the starting oil, and the emulsion will remain stable for weeks and even months after formation. Stable emulsions are reddish-brown in color and appear to be nearly solid. Because of their high viscosity and near solidity, these emulsions do not spread and tend to remain in lumps or mats on the sea or shore.

The fifth way that oil can contain water is by viscosity entrainment. If the viscosity of the oil is such that droplets can penetrate but will only slowly migrate downward, the oil can contain about 30% to 40% water as long as it is

PHOTO 4.4
A laboratory test of Oil-Fines Interaction (OFI): the brown material in the test flasks is oil interacted with clay fine materials. Some of the OFI can be seen to be precipitated to the bottom of the flasks. (Photo courtesy of Ali Khelifa, Environment Canada.)

in an energetic sea. Once the sea calms or the oil is removed, the water slowly drains. Typically most of the water would be gone before about 2 days. Such water uptake is called **entrained water**. This is not an emulsion type such as the mesostable or stable emulsions, which are stabilized by the chemical action of resins and asphaltenes.

The formation of emulsions is an important event in an oil spill. First, and most important, it substantially increases the actual volume of the spill. Emulsions of all types initially contain about 50% to 70% water and thus when emulsions are formed the volume of the oil spill can be more than tripled. Even more significant, the viscosity of the oil increases by as much as 1000 times, depending on the type of emulsion formed. For example, an oil that has the viscosity of a motor oil can triple in volume and become almost solid through the process of emulsification.

These increases in volume and viscosity make cleanup operations more difficult. Stable emulsified oil is difficult or impossible to disperse, to recover with skimmers, or to burn. Emulsions can be broken down with special chemicals in order to recover the oil with skimmers or to burn it. It is thought that emulsions break down into oil and water by further weathering, oxidation, by dilution with unemulsified oils, and freeze–thaw action. Mesostable emulsions are relatively easy to break down, whereas stable emulsions may take months or years to break down naturally.

Emulsion formation also changes the fate of the oil. It has been noted that when oil forms stable or mesostable emulsions, evaporation slows considerably. Biodegradation also slows. The dissolution of soluble components from oil may also cease once emulsification has occurred.

Natural Dispersion

Natural dispersion occurs when fine droplets of oil are transferred into the water column by wave action or turbulence. Small oil droplets (less than 20 μm or 0.020 mm) are relatively stable in water and will remain so for

PHOTO 4.5
Many tar balls line the surf line of this beach. Tar balls are typically highly weathered oils.
(Photo from the U.S. Coast Guard Web site: http://cgvi.uscg.mil.)

long periods of time. Large droplets tend to rise and larger droplets (larger than 100 μm) will not stay in the water column for more than a few seconds. Depending on oil conditions and the amount of sea energy available, natural dispersion can be insignificant or it can remove the bulk of the oil. In 1993, the oil from a stricken ship, the *Braer*, dispersed almost entirely as a result of high seas off Scotland at the time of the spill and the dispersible nature of the oil cargo.

Natural dispersion is dependent on both the oil properties and the amount of sea energy. Heavy oils such as Bunker C or a heavy crude will not disperse naturally to any significant extent, whereas light crudes and diesel fuel can disperse significantly if the saturate content is high and the asphaltene and resin contents are low. In addition, significant wave action is needed to disperse oil. In 30 years of monitoring spills on the oceans, those spills where oil has dispersed naturally have all occurred in very energetic seas.

The long-term fate of dispersed oil is not known, although it probably degrades to some extent as it consists primarily of saturate components.

Some of the dispersed oil may also rise and form another surface slick, or it may become associated with sediment and be precipitated to the bottom.

Dissolution

Through the process of dissolution, some of the most soluble components of the oil are lost to the water under the slick. These include some of the lower molecular weight aromatics and some of the polar compounds, broadly categorized as resins. As only a small amount, usually much less than a fraction of a percent of the oil, actually enters the water column, dissolution does not measurably change the mass balance of the oil in the environment. The significance of dissolution is that the soluble aromatic compounds are particularly toxic to fish and other aquatic life. If a spill of oil containing a large amount of soluble aromatic components occurs in shallow water and creates a high localized concentration of compounds, then significant numbers of aquatic organisms can be killed.

PHOTO 4.6
Highly weathered oil and emulsion floating near a shoreline. (Photo from the U.S. Coast Guard Web site: http://cgvi.uscg.mil.)

Gasoline, diesel fuel, and light crude oils are the most likely to cause aquatic toxicity. A highly weathered oil is unlikely to dissolve into the water. On open water, the concentrations of hydrocarbons in the water column are unlikely to kill aquatic organisms.

Dissolution occurs immediately after the spill occurs and the rate of dissolution decreases rapidly after the spill as soluble substances are quickly depleted. Some of the soluble compounds also rapidly evaporate.

Photooxidation

Photooxidation can change the composition of an oil. It occurs when the sun's action on an oil slick causes oxygen and carbons to combine and form new products that sometimes are like resins. The photooxidized products may be somewhat soluble and dissolve into the water. It is not well understood how photooxidation specifically affects oils, although certain oils are susceptible to the process, whereas others are not. For most oils, photooxidation is not an important process in terms of changing their fate or mass balance after a spill.

Sedimentation, Adhesion to Surfaces, and Oil–Fines Interaction

Sedimentation is the process by which oil is deposited on the bottom of the sea or other water body. Although the process itself is not fully understood, certain facts about it are. Most sedimentation noted in the past has occurred when oil droplets reached a higher density than water after interacting with

PHOTO 4.7

Very small tar balls litter a beach. (Photo courtesy of the U.S. National Oceanic and Atmospheric Administration [NOAA].)

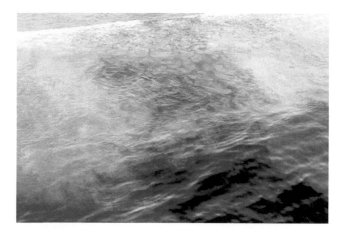

PHOTO 4.8
Waves can have a strong influence on the movement and spreading of oil. In this case, the oil is being spread out by the wave.

mineral matter in the water column. This interaction sometimes occurs on the shoreline or very close to the shore. Once oil is on the bottom, it is usually covered by other sediment and degrades very slowly. In a few well-studied spills, a significant amount (about 10%) of the oil was sedimented on the seafloor. Such amounts can be very harmful to biota that inevitably come in contact with the oil on the sea bottom. Because of the difficulty of studying sedimentation, data are limited.

Oil is very adhesive, especially when it is moderately weathered, and binds to shoreline materials or other mineral material with which it comes in contact. A significant amount of oil can be left in the environment after a spill in the form of residual amounts adhering to shorelines and man-made structures such as piers. As this weathered oil usually contains a high percentage of aromatics and asphaltenes with high molecular weight, it does not degrade significantly and can remain in the environment for decades.

Oil slicks and oil on shorelines sometimes interact with mineral fines suspended in the water column, and the oil is thereby transferred to the water column. Particles of mineral with oil attached may be heavier than water and sink to the bottom as sediment, or the oil may detach and refloat. Oil–fines interaction does not generally play a significant role in the fate of most oil spills in their early stages but can have an impact on the rejuvenation of an oiled shoreline over the long term.

Biodegradation

A large number of microorganisms are capable of degrading petroleum hydrocarbons. Many species of bacteria, fungi, and yeasts metabolize petroleum hydrocarbons as a food energy source. Bacteria and other degrading

PHOTO 4.9
Oil moving away from a stricken cargo vessel along with the winds and currents. In this short distance, little spreading is seen. (Photo courtesy of the International Tanker Owners Pollution Federation [ITOPF].)

organisms are most abundant in areas where there have been petroleum seeps, although these microorganisms are found everywhere in the environment. As each species can utilize only a few related compounds at most, broad-spectrum degradation does not occur. Hydrocarbons metabolized by microorganisms are generally converted to an oxidized compound, which may be further degraded, may be soluble, or may accumulate in the remaining oil. The aquatic toxicity of the biodegradation products is sometimes greater than that of the parent compounds.

The rate of biodegradation depends primarily on the nature of the hydrocarbons and then on the temperature. Generally, rates of degradation tend to increase as the temperature rises. Some groupings of bacteria, however, function better at lower temperatures and others function better at higher temperatures. Indigenous bacteria and other microorganisms are often the best adapted and most effective at degrading oil as they are acclimatized to the temperatures and other conditions of the area. Adding "superbugs" to the oil does not necessarily improve the degradation rate.

The rate of biodegradation is greatest on saturates, particularly those containing approximately 12 to 20 carbons. Aromatics and asphaltenes, which have a high molecular weight, biodegrade very slowly, if at all. This explains the durability of roof shingles containing tar and roads made of asphalt, as both tar and asphalt consist primarily of aromatics and asphaltenes. On the other hand, diesel fuel is a highly biodegradable oil as it is largely composed of biodegradable saturates. Light crude oils are also biodegradable to a degree. Although gasoline contains biodegradable components, it also contains some compounds that are toxic to some microorganisms. These compounds generally evaporate rapidly. Gasoline will generally evaporate before it can degrade. Heavy crude oils contain little material that is readily biodegradable and Bunker C contains almost none.

The rate of biodegradation is also highly dependent on the availability of oxygen. On land, oils such as diesel can degrade rapidly at the surface, but very slowly if at all only a few centimeters below the surface, depending on oxygen availability. In water, oxygen levels can be so low that degradation is limited. It is estimated that it would take all the dissolved oxygen in approximately 400,000 L of seawater to completely degrade 1 L of oil. The rate of degradation also depends on the availability of nutrients such as nitrogen and phosphorus, which are most likely to be available on shorelines or on land. Finally, the rate of biodegradation also depends on the availability of the oil to the bacteria or microorganism. Oil degrades significantly at the oil–water interface at sea, and on land mostly at the interface between soil, oil, and air.

Biodegradation can be a very slow process for some oils. It may take weeks for 50% of a diesel fuel to biodegrade under optimal conditions and years for 10% of a crude oil to biodegrade under less optimal conditions. For this reason, biodegradation is not considered an important weathering process in the short term.

Formation of Tar Balls

Tar balls are agglomerations of heavy oil less than about 10 cm in diameter. Larger accumulations of the same material ranging from about 10 cm to 1 m in diameter are called tar mats. Tar mats are pancake shaped rather than round. Their formation is still not completely understood, but it is known that they are formed from the residuals of heavy crudes and Bunker C. After these oils weather at sea and slicks are broken up, the residuals remain in

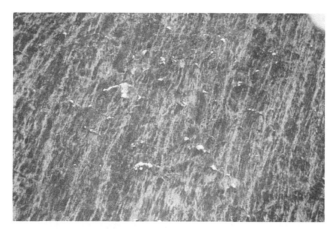

PHOTO 4.10
An aerial view of sheen strung out in windrows. At sea, oil often is found in such windrows.

PHOTO 4.11
Another aerial view of heavy oil spread out in windrows.

tar balls or tar mats. The formation of droplets into tar balls and tar mats has also been observed, with the binding force being simply adhesion.

The formation of tar balls is the ultimate fate of many spilled oils at sea. These tar balls are then deposited on shorelines around the world. The oil may come from spills, but it is also residual oil from natural oil seeps or from deliberate operational releases such as from ships. Tar balls are regularly recovered by machine or by hand from recreational beaches by government agencies or resort owners.

Movement of Oil and Oil Spill Modeling

Spreading

The spreading of oil spilled on water is discussed in this section. Oil spreads to a lesser extent and very slowly on land than on water. The spreading of oil on land is described in Chapter 12. Oil spilled on or under ice spreads relatively rapidly but does not spread to as thin a slick as on water. On any surface other than water, such as ice or land, a large amount of oil is retained in depressions, cracks, and other surface irregularities.

After an oil spill on water, the oil tends to spread into a slick over the water surface. This is especially true of the lighter products such as gasoline, diesel fuel, and light crude oils, which form very thin slicks. Heavier crudes and Bunker C spread to slicks as much as several millimeters thick. Heavy oils may also form tar balls and tar mats and thus may not go through progressive stages of thinning. The area of spreading for these different types of oil is illustrated in Figure 4.2.

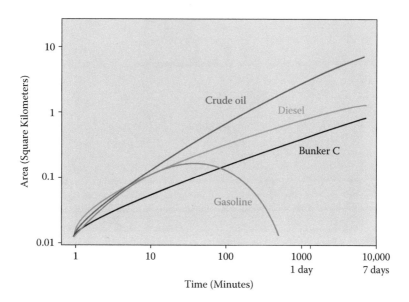

FIGURE 4.2
Comparison of the spreading rates of different oils and fuels.

Oil spreads horizontally over the water surface even in the complete absence of wind and water currents. This spreading is caused by the force of gravity and the interfacial tension between oil and water. The viscosity of the oil opposes these forces. As time passes, the effect of gravity on the oil diminishes, but the force of the interfacial tension continues to spread the oil. The transition between these forces takes place in the first few hours after the spill occurs.

The rates of spreading under ideal conditions are shown in Figure 4.2. As a general rule, an oil slick on water spreads relatively quickly immediately after a spill. The outer edges of a typical slick are usually thinner than the inside of the slick at this stage so that the slick may resemble a fried egg. After a day or so of spreading, this effect diminishes.

Winds and currents also spread the oil and speed up the process. Oil slicks will elongate in the direction of the wind and currents, and as spreading progresses, take on many shapes depending on the driving forces. Oil sheens often precede heavier or thicker oil concentrations. If the winds are high (more than 20 km/h), the sheen may separate from thicker slicks and move downwind.

A slick often breaks into "windrows" on the sea under the influence of either waves or zones of convergence or divergence. Oil tends to concentrate between the crests of waves simply due to the force of gravity. There are often vertical circulation cells in the top 20 m of the sea. When two circulation cells meet, a zone of convergence is formed. When two currents diverge, it forms a zone of divergence. Oil moving along these zones is alternately concentrated and spread by the circulation currents to form ribbons

PHOTO 4.12
Oil spreading out from an undersea blowout. (Photo from the U.S. Coast Guard Web site:
http://cgvi.uscg.mil.)

or windrows of oil rather than continuous slicks. In some locations close
to shore, zones of convergence and divergence often occur in similar locations so that oil spills may appear to have similar trajectories and spreading
behavior in these areas.

Movement of Oil Slicks

In addition to their natural tendency to spread, oil slicks on water are
moved along the water surface, primarily by surface currents and winds.
The slick generally moves at a rate that is 100% of the surface current and
approximately 3% of the wind speed. If the wind is more than about 20
km/h, however, and the slick is on the open sea, wind predominates in
determining the slick's movement. Both the wind and surface current must
be considered for most situations. This type of movement is illustrated
in Figure 4.3.

When attempting to determine the movement of an oil slick, two factors
affect accuracy. The more significant factor is the inability to obtain accurate
wind and current speeds at the time of a spill. The other, very minor factor is a phenomenon commonly known as the Coriolis effect, whereby the
earth's rotation deflects a moving object slightly to the right in the Northern
Hemisphere and to the left in the Southern Hemisphere.

Sinking and Overwashing

If oil is denser than the surface water, it may sometimes actually sink. Some
rare types of heavy crudes and Bunker C can reach these densities and sink.
When this occurs, the oil may sink to a denser layer of water rather than to the

PHOTO 4.13
Overwashed oil residue from an in-situ burn (reddish in appearance). The thin layer of water on this heavy residue prevents one from seeing the oil except from nearby and at a high angle.

PHOTO 4.14
Ice can change the fate and behavior of oil spills. In this photo, Bunker oil has been broken up into small particles by the ice and contaminates a wide area of ice surface. When the ice melted, much of this oil refloated and contaminated a nearby shoreline.

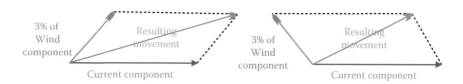

FIGURE 4.3
The effect of different wind and current directions on the movement of an oil slick.

PHOTO 4.15
Oil spreading on top of the water from a sunken World War II ship. The oil begins arriving at the surface in the middle of the photo and then spreads out and moves with the wind and currents. (Photo courtesy of the U.S. National Oceanic and Atmospheric Administration [NOAA].)

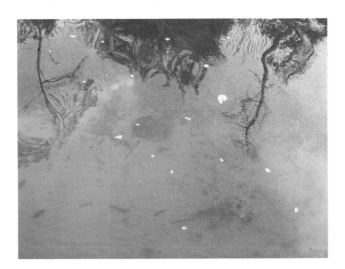

PHOTO 4.16
Oil patties sunken on the bottom of a lake. The red arrow points to one of the oil patties. The water is about 1 meter deep.

bottom. Less dense layers of water may override these denser layers of water. This occurs, for example, when seas are not high and warmer freshwater from land overrides dense seawater. The freshwater may have a density of about 1.00 g/mL and the seawater a density of about 1.03 g/mL. An oil with a density greater than 1.00 but less than 1.03 would sink through the layer of freshwater and ride on the layer of saltwater. The layer of freshwater usually varies in depth from about 1 to 10 m. If the sea energy increases, the oil may actually reappear on the surface, as mixing increases the density of the water from 1.00 to about 1.03.

It is important to note that sinking of any form, whether to the bottom or to the top of a layer of dense seawater, is rare. When oil does sink, it complicates cleanup operations as the oil can be recovered only with specialized underwater suction devices or special dredges.

Overwashing is another phenomena that occurs quite frequently and can hamper cleanup efforts. At moderate sea states, a dense slick can be overwashed with water. When this occurs, the oil can disappear from view if the spill is being observed from an oblique angle, as would occur if someone is looking away from a ship. Overwashing causes confusion about the fate of an oil spill, as it can give the impression that the oil has sunk and then resurfaced.

Spill Modeling

Spill response personnel need to know the direction in which an oil spill is moving in order to protect sensitive resources and coastline. To assist them with this, computerized mathematical models have been developed to predict the trajectory or pathway and fate of oil. Outputs of one such spill model are shown in Figure 4.4.

Today's sophisticated spill models combine the latest information on oil fate and behavior with computer technology to predict where the oil will go and what state it will be in when it gets there. Their major limitation to accurately predicting an oil slick's movement is the lack of accurate estimates of water current and wind speeds along the predicted path. This is likely to remain a limitation in the future.

In addition to predicting the trajectory, these models can estimate the amount of evaporation, the possibility of emulsification, the amount of dissolution and the trajectory of the dissolved component, the amount and trajectory of the portion that is naturally dispersed, and the amount of oil deposited and remaining on shorelines. Accurate spill modeling is now a very important part of both contingency planning and actual spill response.

Spill models operate in a variety of modes. The most typical is the trajectory mode, which predicts the trajectory and weathering of the oil. The stochastic mode uses available data to predict a variety of scenarios for the oil spill, which includes the direction, fate, and property changes in the oil slick. In another mode, often called the receptor mode, a site on the shore or water is chosen and the trajectory from the source of the oil is calculated. Increasingly, statistically generated estimates are added to oil spill models to compensate for the lack of immediate data on winds and currents.

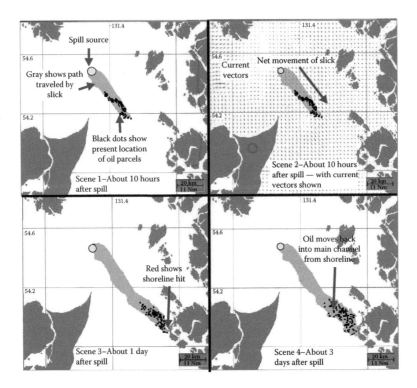

FIGURE 4.4
Example outputs from a typical spill trajectory model.

PHOTO 4.17
Oil spreads along a shoreline and moves with the wind and current from a Bunker C spill on a lake.

5

Analysis, Detection, and Remote Sensing of Oil Spills

Special instruments are sometimes required to detect an oil spill, especially if the slick is very thin or not clearly visible. For example, if a spill occurs at night, in ice, or among seaweeds, the oil slick must be detected and tracked using instruments onboard aircraft or satellites. This technology is known as remote sensing. There are also surface technologies available to detect and track oil slicks. In addition, samples of the oil must often be obtained and analyzed to determine the oil's properties, its degree of weathering, its source, or its potential impact on the environment. This analysis as well as tracking and remote sensing technologies are discussed in this chapter.

The Importance of Analytical and Detection Technologies

Technology is currently available to give the location and extent of the spill, the potential behavior of the oil, and its impact on the environment. Laboratory analysis can provide information to help identify an oil if its source is unknown and a sample is available. With a sample of the source oil, the degree of weathering and the amount of evaporation or biodegradation can be determined for the spilled oil. Through laboratory analysis, the more toxic compounds in the oil can be measured and the relative toxicity of the oil at various stages of the spill can be determined. This is important information to have as the spill progresses.

Sampling and Laboratory Analysis

Taking a sample of oil and then transporting it to a laboratory for subsequent analysis is common practice. Although there are many procedures for taking oil samples, it is always important to ensure that the oil is not tainted from contact with other materials and that the sample bottles are precleaned with solvents, such as hexane, that are suitable for the oil.

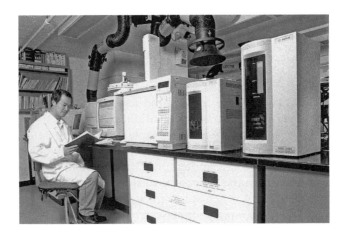

The simplest and most common form of analysis is to measure how much oil is in water, soil, or sediment sample. Such analysis results in a value known as **total petroleum hydrocarbons (TPHs)**. The TPH measurement can be obtained in many ways, including extracting the soil or evaporating a solvent such as hexane and measuring the weight of the residue, which is presumed to be oil.

The oil can also be extracted from water using an oil-absorbing solid. The oil is then analyzed from this substrate by a variety of means. Still another method is to use enzymes that are selectively affected by some of the oil's components. Test kits are available, which use color to indicate the effect of the oil on the enzymes.

A more sophisticated form of analysis is to use a gas chromatograph (GC). A small sample of the oil extract, often in hexane, and a carrier gas, usually helium, are passed through a small glass capillary. The glass column is coated with absorbing materials and, as the various components of the oil have varying rates of adhesion, the oil separates as these components are absorbed at different rates onto the column walls and subsequently evaporated as the temperature increases. The gases then pass through a sensitive detector. The system is calibrated by passing known amounts of standard materials through the unit. The amount of many individual components in the oil is thereby measured. The components that pass through the detector can also be totaled and a TPH value determined. Although it is highly accurate, this TPH value does not include resins, asphaltenes, and some other components of the oil with higher molecular weight that do not pass through the column. Typical chromatograms of crude oils with some of the more prominent components of the oil identified are shown in Figure 5.1.

One type of detector used on a gas chromatogram is a mass spectrometer (MS). The method is usually called GC-MS and can be used to quantify and identify many components in oil. The mass spectrometer provides information about the structure of the substance so that each peak in the chromatogram can be more positively identified. This information can then be used to predict how long the oil has been in the environment and what percentage of it has evaporated or biodegraded. This is possible because some of the components in oils, particularly crude oils, are very resistant to biodegradation, whereas others are resistant to evaporation. This difference in the distribution of components then allows the degree of weathering of the oil to be measured. The same technique can be used to "fingerprint" an oil and positively identify its source. Certain compounds are consistently distributed in oil, regardless of weathering, and these are used to identify the specific type of oil.

Field Analysis

Analysis performed in the field is faster and more economical than analysis done in a laboratory. As analytical techniques are constantly improving and lighter and more portable equipment is being developed, more analytical

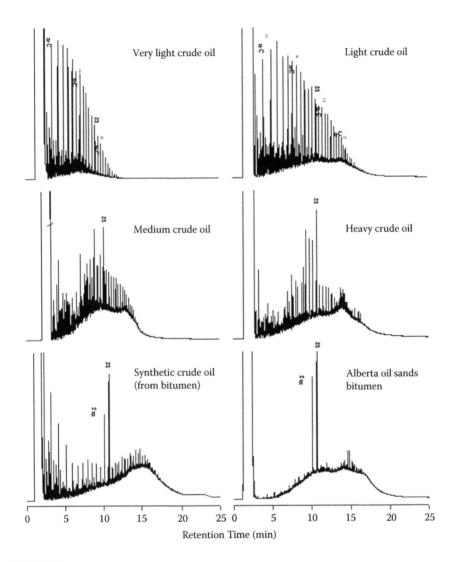

FIGURE 5.1
Chromatograms of several types of oil.

work can be carried out directly in the field. Test methods are now available for measuring physical properties of oil such as viscosity, density, and even flash point in the field. Test kits have also been developed that can measure total petroleum hydrocarbons directly in the field. Although these test kits are less accurate than laboratory methods, they are rapid screening tools that minimize time for analysis and may provide adequate data for making response decisions.

PHOTO 5.3
Two environmental officers taking samples of the oil from a train derailment. (Photo courtesy of Environment Canada.)

Detection and Surveillance

Detection and Tracking Buoys and Systems

As oil spills frequently occur at moorings and docks, buoys and fixed point monitoring systems have been developed to ensure rapid response at these sites. These systems detect the oil on water and transmit a radio signal to an oil spill response agency. Fluorescence is one method used to detect oil in these systems. An ultraviolet light is focused on the water surface and any oil that is present fluoresces, or absorbs the ultraviolet light and reemits it as visible light. This fluorescing phenomenon is relatively unique to oil and provides a positive detection mechanism.

As an oil spill moves with the winds and surface currents, the slick or portions of it may move and responders may not always know its position, especially in darkness or fog. Buoys have been developed that move on the water in a manner similar to oil. These buoys transmit a position signal directly to receivers located on aircraft or ships or to a satellite that corresponds to the position of the oil slick. Some of these buoys receive Global Positioning System (GPS) data from satellites and transmit this with the signal. The position of the spill can then be determined using a remote receiver. For this type of device to be effective, however, the buoy must respond to both the wind and surface currents in the same way as the oil would. Although this precision in response is difficult to achieve, devices are available that can successfully track a range of crude oils and petroleum products.

PHOTO 5.4
A field technician is lowering a field fluorometer to estimate oil-in-water concentrations. (Photo from the U.S. Coast Guard Web site: http://cgvi.uscg.mil.)

PHOTO 5.5
A number of tracking and sampling buoys undergoing tests.

Visual Surveillance

Oil spills are often located and surveyed from helicopters or aircraft using only human vision. There are some conditions, however, such as fog and darkness, in which oil on the surface cannot be seen. Very thin oil sheens are also difficult to detect as is oil viewed from an oblique angle (less than 45°) especially in misty or other conditions that limit vision. Oil can also be difficult to see in high seas and among debris or weeds, and it can blend into dark backgrounds, such as water, soil, or shorelines.

In addition, many naturally occurring substances or phenomena can be mistaken for oil. These include weeds and sunken kelp beds, whale and fish sperm, biogenic or natural oils such as from plants, glacial flour (finely ground mineral material), sea spume (organic material), wave shadows, sun glint and wind sheens on water, and oceanic and riverine fronts where two different bodies of water meet, such as a river entering the sea.

A very thin oil sheen as it appears on water is shown in Figure 5.2. This figure also shows the thickness and amount of oil that could be present

Approximate Slick Thickness (µm)	Oil Appearance	Photographic Example
0.05 to 0.2	Silvery sheen	
0.3 to 3	Rainbow sheen	
>3	Oil colored brown to black	

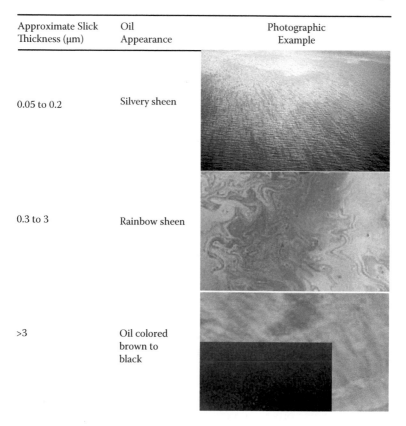

FIGURE 5.2
The appearance of thin films of oil on water.

under such circumstances. It is important to note that color and slick appearance are only roughly related to thickness and that this relationship only applies to very thin slicks.

Remote Sensing

Remote sensing of oil involves the use of sensors other than human vision to detect or map oil spills. As already noted, oil often cannot be detected in certain conditions. Remote sensing provides a timely means to map out the locations and approximate concentrations of very large spills in many conditions. Remote sensing is usually carried out with instruments onboard aircraft or from satellite. Although many sensors have been developed for a variety of environmental applications, only a few are useful for oil spill work. Remote sensing of oil on land is particularly limited and only one or two sensors are useful. Table 5.1 shows a summary of sensors, their characteristics, and their suitability to various missions.

Visual Sensors

Many devices employing the visible spectrum, including the conventional camera and video camera, are available at a low cost. As these devices are subject to the same interferences as visual surveillance, they are used primarily to document the spill or to provide a frame of reference for other sensors.

Infrared Sensors

Thick oil on water absorbs infrared radiation from the sun and thus appears in infrared imagery as hot on a cold ocean surface. Unfortunately, many other false targets such as weeds, biogenic oils, debris, and oceanic and riverine fronts can interfere with oil detection. The advantage of infrared sensors over visual sensors is that they give information about relative thickness since only thicker slicks, probably greater than about 50 μm, show up in the infrared.

Infrared imagery also may be of some use at night when the oil appears "colder" than the surrounding sea. Often oil is not detectable at night in the infrared for a variety of interfering conditions.

Infrared sensors are relatively inexpensive and widely used for supporting cleanup operations and directing cleanup crews to thicker portions of an oil spill. They are also often used on cleanup vessels. The oblique view from a ship's mast is often sufficient to provide useful information on where to steer the vessel for best oil recovery over a short range.

TABLE 5.1

Sensor Characteristics and Suitability for Various Missions

| Sensor | Mission Suitability | | | | | | Typical Coverage Width (km) | Acquisition Cost Range (thousands $) | State of Development |
	Support for Cleanup	Night and Fog Operation	Oiled Shoreline Survey	Spill Mapping	Ship Discharge Surveillance			
Still camera	2	n/a	2	2	2	0.25 to 2	1 to 5	High
Video	2	n/a	2	2	2	0.25 to 5	1 to 10	High
Nighttime vision camera	3	4	n/a	2	2	0.25 to 2	5 to 20	Medium
IR camera (8–14 µm)	4	2	n/a	3	3	0.25 to 2	20 to 50	High
UV camera	2	n/a	n/a	3	2	0.25 to 2	4 to 20	Medium
Multispectral scanner	1	n/a	1	2	1	0.25 to 2	100 to 200	Medium
Radar	n/a	4	n/a	4	3	5 to 50	1200 to 8000	High
Microwave radiometer	1	3	n/a	2	2	1 to 5	400 to 1000	Medium
Laser fluorosensor	4	3	5	1	5	0.01 to 0.1	300 to 1000	Medium

Note: n/a, not applicable. Numerical values represent a scale from 1 = poorly suited to 5 = ideally suited.

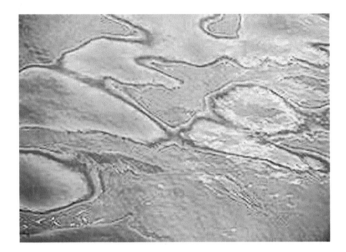

PHOTO 5.6
The only oil color that is strongly based in physics is that of a rainbow sheen, as shown here. This color occurs rarely and represents a very thin sheen. Appearance and color of oil can be misleading as an indication of thickness.

PHOTO 5.7
What appears to be oil in this photograph is a riverfront protruding into the ocean. This highlights the need for specialized remote sensing systems.

Laser Fluorosensors

Oils that contain aromatic compounds, as most oils do, will absorb ultraviolet light and give off visible light in response. Since very few other compounds respond in this way, this can be used as a positive method of detecting oil at sea or on land. Laser fluorosensors use a laser in the ultraviolet spectrum to

PHOTO 5.8
A visual image of a slick. Sun glint causes the image to be unclear.

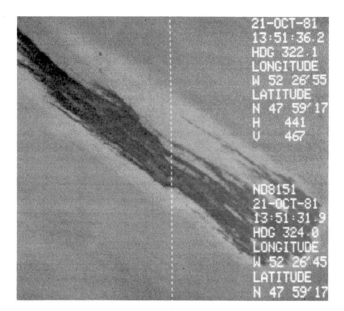

PHOTO 5.9
The same image as in Photo 5.8, but using infrared. In this case, the infrared technique produced a much clearer image.

trigger this fluorescing phenomenon and a sensitive light detection system to provide an oil-specific detection tool. There is also some information in the visible light return that can be used to determine whether the oil is a light or heavy oil or a lubricating oil. In a sense, using laser fluorosensors is like performing chemistry from the air. Laser fluorosensors are the most powerful remote sensing tools available because they are subject to very few interferences. Laser fluorosensors work equally well on water and on land, and are the only reliable means of detecting oil in certain ice and snow situations. Disadvantages include the high cost of these sensors and their large size and weight.

Passive Microwave Sensors

The passive microwave sensor detects natural background microwave radiation. Oil slicks on water absorb some of this signal in proportion to their thickness. Although this cannot be used to measure thickness absolutely, it can yield a measure of relative thickness. The advantage of this sensor is that it can detect oil through fog and in darkness. The disadvantages are the poor spatial resolution and relatively high cost.

Thickness Sensors

Some types of sensors can be used to measure the thickness of an oil slick. For example, the passive microwave sensor can be calibrated to measure the relative thickness of an oil slick. Absolute thickness cannot be measured for the following reasons: many other factors such as atmospheric conditions also change the radiation levels; the signal changes in cyclical fashion with spill thickness; and the signal must be averaged over a relatively wide area and the slick can change throughout this area. The cyclical nature of the microwave radiation variance with thickness has been overcome by some suppliers by using multifrequency units, for example, five different frequencies.

The infrared sensor also measures only relative thickness. Thicker oil appears hotter than the surrounding water during daytime. Although the degree of brightness of the infrared signal changes little with thickness, some systems have been adjusted to yield two levels of thickness. In any case, the onset of an infrared signal from an oil spill represents a slick that is just barely thicker than a sheen and does not give enough information to assist the spill responder other than this slick is not a sheen.

Sensors using lasers to create sound waves through oil can measure absolute oil thickness. The time it takes the sound waves to travel through the oil changes little with the type of oil and thus the measurement of this travel time yields a reliable measurement of the oil's thickness. This type of sensor is still considered experimental.

Radar

As oil on the sea calms smaller waves (waves on the order of a few centimeters in length), radar can detect oil on the sea as a calm area. The technique is highly prone to false targets, however, and is limited to a narrow range of wind speeds (approximately 2 to 6 m/s). At winds below this wind speed, there are not enough small waves to yield a difference between the oiled area and the sea. At higher winds, the waves can propagate through the oil and the radar may not be able to "see" into the troughs between the waves. Radar is not useful near coastlines or between headlands because the wind "shadows" look like oil. There are also many natural calms on the oceans that can resemble oil. Despite its large size and expense, radar equipment is particularly well suited for searches of large areas and for work at night or in foggy or other bad weather conditions.

Satellites

Although many satellites provide images in the visible spectrum, oil cannot be seen in these images unless the spill is very large or rare sea conditions are prevalent that provide a contrast to the oil. Oil has no spectral characteristics that allow it to be enhanced from the background.

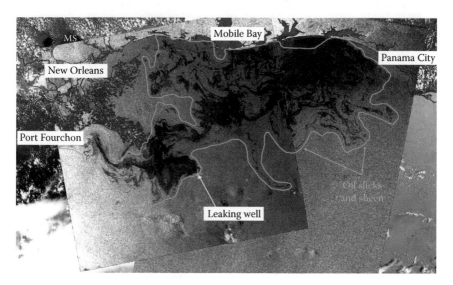

PHOTO 5.10
A wide area coverage of the Gulf of Mexico during the 2010 Macondo well blowout. The radar image is placed on a visible satellite image (colored) to give it a spatial reference. The slick is outlined by the orange line. Close to shore, radar does not provide useful oil detection and this is shown by the placement of the orange line to areas of greater probability. This overview of spills can be quite useful for the responder. (Photo courtesy of Radarsat.)

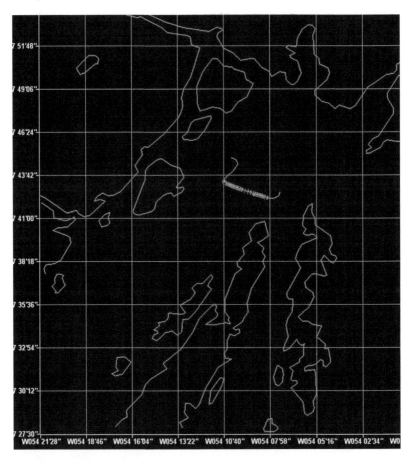

PHOTO 5.11
A map-like display from a laser fluorosensor instrument. The red shows a positive oil detection of a given type. The blue line shows the flight path. (Photo courtesy of Environment Canada.)

PHOTO 5.12
A remote sensing airplane. This aircraft contains an experimental radar system as well as other remote sensing equipment. The radar antennae are in the black bulbs protruding from the airplane. (Photo courtesy of Environment Canada.)

Several radar satellites are available that operate in the same manner as airborne radar and share their many limitations (such as wind shadows from coastlines and limits of wind speeds between 2 to about 6 m/s). Despite these limitations, radar imagery from satellite is particularly useful for mapping large oil spills. Arrangements are possible to provide the data within a few hours, making this a useful option. Much modern spill response depends on radar satellite data.

6

Containment on Water

Containment of an oil spill refers to the process of confining the oil, either to prevent it from spreading to a particular area, to divert it to another area where it can be recovered or treated, or to concentrate the oil so it can be recovered or burned.

Containment booms, or simply booms, are the basic and most frequently used piece of equipment for containing an oil spill on water. Booms are generally the first equipment mobilized at a spill and are often used throughout the operation. This chapter covers the types of booms, their construction, and operating principles and uses, as well as how and why they fail. It also covers ancillary equipment used with booms, sorbent booms, and special purpose and improvised booms. The topic of fire-resistant booms for use when burning oil on water is covered in Chapter 10.

Types of Booms and Their Construction

A boom is a floating mechanical barrier designed to stop or divert the movement of oil on water. Booms resemble a vertical curtain with portions extending above and below the waterline. Most commercial booms consist of four basic components: a means of flotation, a freeboard member (or section) to prevent oil from flowing over the top of the boom, a skirt to prevent oil from being swept underneath the boom, and one or more tension members to support the entire boom. Booms are constructed in sections, usually 15 or 30 m long, with connectors installed on each end so that sections of the boom can be attached to each other, towed, or anchored. A section of a typical boom is shown in Figure 6.1. Some typical boom types are illustrated in Figure 6.2.

The *flotation members* or *floats* determine the buoyancy of the boom and keep it floating on the water surface. They are located along the center line; outboard, on one side; or on outriggers. Booms either have solid floats or the boom itself is inflatable. Solid floats are usually made of a plastic foam, such as expanded polyurethane or polyethylene, and are segmented or flexible so the boom can ride the surface of the waves. Inflatable booms are either self-inflating or are inflated using a powered air source. They require little storage space but are generally less rugged than booms with fixed floats.

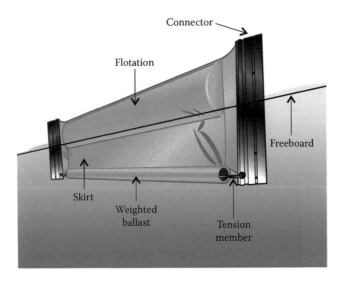

FIGURE 6.1
Basic boom construction.

The **freeboard** member is the portion of the boom above the water that prevents oil from washing over the top of the boom. The term *freeboard* itself is also used to refer to the height from the water line to the top of the boom. The *skirt* is the portion of the boom below the floats or flotation that helps to contain the oil. It is usually made of the same types of fabric as the freeboard member and the covering of the floats. Boom fabrics consist of various fiber types such as nylon, polyester, aramid (Kevlar®), or blends of these. These are then saturated or coated with polymeric coatings of various types ranging from polyvinyl chloride (PVC), polyurethane, nitrile, or blends of these materials to provide an impervious layer over the woven base fabric. Fabrics are designed for a range of operating conditions based upon the desired lifetime, sun and salt resistance, and abrasion.

Most booms are also fitted with one or more **tension members** that run along the bottom of the boom and reinforce it against the horizontal load imposed by waves and currents. Tension members are usually made of steel cables or chains but sometimes consist of nylon or polyester ropes. The boom fabric itself is not strong enough to withstand the powerful forces to which booms are subjected, except in protected waters. For example, the force on a 100-m long section of boom could be as much as 10,000 kg, depending on sea conditions and the construction of the boom.

Booms are sometimes constructed with *ballast* or weights designed to maintain the boom in an upright position. Lead weights have been used for this, but steel chain in the bottom of the boom often serves as both ballast and tension member. A few booms also use a chamber filled with water as ballast. Many booms nowadays are constructed without ballast, however, and their position

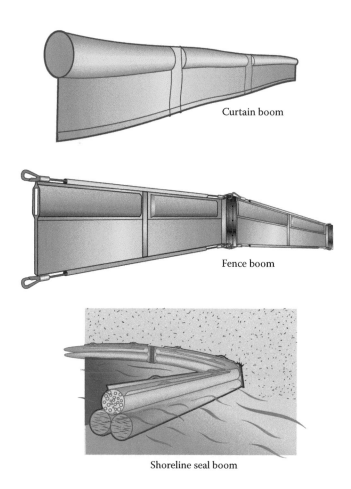

Curtain boom

Fence boom

Shoreline seal boom

FIGURE 6.2
Three common boom types.

in the water is maintained by balancing the forces on the top and bottom of the boom. Another construction feature common in larger booms is the addition of "stiffeners" or rigid strips, often consisting of plastic, aluminum or steel bars, which are designed to support the boom and keep it in an upright position.

The three basic types of booms are (1) fence and (2) curtain booms, which are common, and (3) shoreline seal booms, which are relatively rare. Booms are also classified according to where they are used, that is, offshore, inshore, harbor, and river booms, based on their size and ruggedness of construction. The *fence boom* is constructed with a freeboard member above the float. Although relatively inexpensive, these booms are not recommended for use in high winds or strong water currents. *Curtain booms* are constructed with a skirt below the floats and no freeboard member above the float. Curtain booms are most suitable for use in strong water currents. *Shoreline seal booms*, which are generally constructed

PHOTO 6.1
A series of booms have been placed around a leaking ship. (Photo courtesy of Environment
Canada.)

PHOTO 6.2
A permanent harbor boom has been placed near pilings. (Photo courtesy of Aqua-Guard Spill
Response Inc.)

with three tubes, two of which are weighted, are used in sealing shorelines against oil contamination. These three types are illustrated in Figure 6.2.

The characteristics of booms that are important in determining their operating ability are the buoyancy-to-weight ratio or reserve buoyancy, the heave response, and the roll response. The *buoyancy-to-weight ratio* or *reserve buoyancy* is determined by the amount of flotation and the weight of the boom. This means that the float must provide enough buoyancy to balance the weight of the boom with the force exerted by currents and waves, thereby maintaining the boom's stability. The greater a boom's reserve buoyancy, the greater its ability to rise and fall with the waves and remain on the surface of the water. The *heave response* is the boom's ability to conform to sharp waves. It is indicated by the reserve buoyancy and the flexibility of the boom. A boom with a good heave response will move with the waves on the surface of the water and not be alternately submerged and thrust out of the water by the wave action. The *roll response* refers to the boom's ability to remain upright in the water and not roll over.

Uses of Booms

Booms are used to enclose oil and prevent it from spreading; to protect harbors, bays, and biologically sensitive areas; to divert oil to areas where it can be recovered or treated; and to concentrate oil and maintain an even thickness so that skimmers can be used or other cleanup techniques, such as in-situ burning, can be applied.

Booms are used primarily to contain oil, although they are also used to deflect oil. When used for containment, booms are often arranged in a U, V, or J configuration. The U configuration is the most common and is achieved by towing the boom behind two vessels, anchoring the boom, or by combining these two techniques. The U shape is created by the current pushing against the center of the boom. The critical requirement is that the current in the apex of the U does not exceed 0.5 m/s or 1 knot, which is referred to as the **critical velocity**. This is the speed of the current flowing perpendicular to the boom, above which oil will be lost from the boom.

In open water, the U configuration can also be achieved by allowing the entire boom system to move down current so that the velocity of the current, as opposed to that of the boom, does not exceed the critical velocity. If this velocity is exceeded, first small amounts of oil and then massive amounts will be lost. This leads to several types of boom failure, which are described in the next section.

If used in areas where the currents are likely to exceed 0.5 m/s or 1 knot, such as in rivers and estuaries, booms are often used in the deflection mode. The boom is then deployed at various angles to the current, shown in Table 6.1, so that the critical velocity is not exceeded. The oil can then be deflected to areas where it can be collected or to less sensitive areas as shown in Figure 6.3.

If strong currents prevent the best positioning of the boom in relation to the current, several booms can be deployed in a cascading pattern to progressively move oil toward one side of the watercourse. This technique is effective in wide rivers or where strong currents may cause a single boom to fail. The deflection is intended to be in a straight line, but usually cusps form in the boom as a result of the current. When booms are used for deflection, the forces of the current on the boom are usually so powerful that stronger booms are required and they must be anchored along their entire length.

TABLE 6.1

Boom Deflection Angles and Critical Current Velocities

Angle	Velocity of Perpendicular Current before Critical Velocity Is Reached[a]
90	0.5
75	0.5
60	0.6
45	0.7
35	0.9
15	1.9

[a] The velocity of current that would be encountered if the boom was perpendicular to the current. Velocity in meters per second (m/s; is approximately half of current in knots).

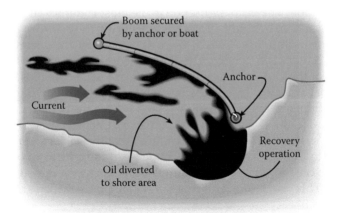

FIGURE 6.3
Using a boom for deflection.

Figure 6.4 shows a variety of boom deployment configurations. The U configuration is also used to keep oil from spreading into bays or other sensitive areas, as well as to collect oil so cleanup measures can be applied. The J configuration is a variation of the U configuration and is usually used to contain oil as well as to deflect it to the containment area. The U and J configurations are easily interchanged. The V configuration usually consists of two booms with a counterforce such as a skimmer at the apex of the two booms.

Encirclement is another way that booms can be used for containment. Stricken ships in shallow waters are often encircled or surrounded by booms to prevent further movement of oil away from the ship. Oil losses usually still occur because the boom's capacity is exceeded or strong currents may sweep the oil under the boom. In many cases, however, this is all that can be done to prevent further spillage and spreading of the oil. Encirclement is often used as a preventative measure at tanker loading and unloading facilities. Because these facilities are usually situated in calm waters, small amounts of oil from minor spills can often be contained using this technique.

Booms are also used in fixed systems attached to docks, piers, harbor walls, or other permanent structures with sliding-type connectors that allow the boom to move up and down with the waves and tide. Their purpose is to protect certain areas from an oil spill. They are also used to enclose an area where oil is frequently loaded or unloaded, or to provide backup containment for operations such as oil–water separators on shore. As these booms are often in place for 10 years before replacement, special long-lasting booms are used.

Booms are also used in a "sweep" configuration to either deflect oil or contain it for pickup by skimmers. The sweep is held away from the vessel by a fixed arm and the boom allowed to form a U shape, as shown in Figure 6.4. A skimmer is usually placed in the U or is sometimes fixed in the vessel's hull and the oil is deflected to this position. Special vessels are required that can maneuver while moving slowly so that the boom does not fail. The various configurations in which booms can be deployed are shown in Figure 6.4.

Boom Failures

Water currents, waves, and winds affect a boom's performance and its ability to contain oil. Either alone or in combination, these forces often lead to boom failure and loss of oil. Eight common ways in which booms fail are discussed here. Some of these are illustrated in Figure 6.5.

> *Entrainment failure*—This type of failure is caused by the speed of the water current and is more likely to happen with a lighter oil. When oil is being contained by a boom in moving water, if the current is fast enough, the boom acts like a dam and the surface water being

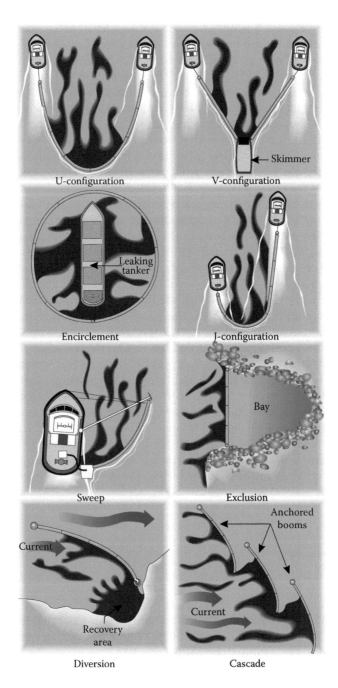

FIGURE 6.4
Configurations for boom deployment.

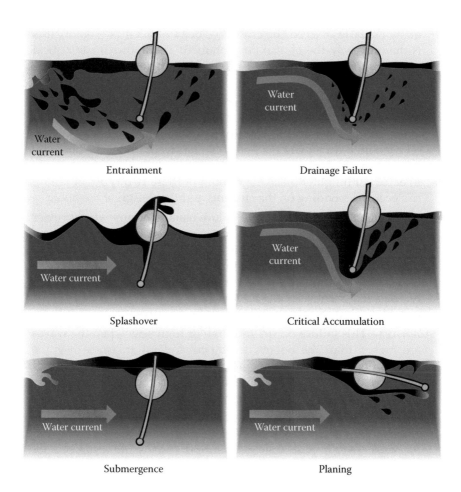

FIGURE 6.5
Boom failure modes.

held back is diverted downward and accelerates in an attempt to keep up with the water flowing directly under the boom. The resulting turbulence causes droplets to break away from the oil that has built up in front of the boom, referred to as the oil headwave, pass under the boom, and resurface behind it. The water speed at which the headwave becomes unstable and the oil droplets begin to break away is referred to as the critical velocity. It is the speed of the current flowing perpendicular to the boom, above which oil losses occur. For most booms riding perpendicular to the current, this critical velocity is about 0.5 m/s (about 1 knot).

At current speeds greater than the critical velocity, this type of boom failure can be overcome by placing the boom at an angle to the current or in the deflection mode. Since currents in most rivers and many estuaries exceed the critical velocity of 0.5 m/s (1 knot), this is the only way the oil can be contained. The approximate critical velocities for booms riding at various other angles to the current are listed in Table 6.1.

Drainage failure—Similar to entrainment, this type of failure is related to the speed of the water current, except that it affects the oil directly at the boom. After critical velocity is reached, large amounts of the oil contained directly at the boom can be swept under the boom by the current. Both entrainment and drainage failure are more likely to occur with lighter oils. One or both of these two types of failure can occur, depending on the currents and the design of the boom.

Critical accumulation—This type of failure usually occurs when heavier oils, which are not likely to become entrained in water, are being contained. Heavier oils tend to accumulate close to the leading edge of the boom and are swept underneath the boom when a certain critical accumulation point occurs. This accumulation is often reached at current velocities approaching the critical velocities listed in Table 6.1 but can also be reached at lower current velocities.

Splashover—This failure occurs in rough or high seas when the waves are higher than the boom's freeboard and oil splashes over the boom's float or freeboard member. It can also occur as a result of extensive oil accumulation in the boom compared to the freeboard.

Submergence failure—This type of failure occurs when water goes over the boom. Often the boom is not buoyant enough to follow the wave motion and some of the boom sinks below the waterline and oil passes over it. Submergence failure is usually the result of poor

PHOTO 6.3
Pieces of a broken boom lying on a beach. This happens when a light duty boom encounters higher seas. (Photo from the U.S. Coast Guard Web site: http://cgvi.uscg.mil.)

heave response, which is measured by both the reserve buoyancy and the flexibility of the boom. Failure due to submergence is not that common as other forms of failure, such as entrainment, usually occur first.

Planing—Planing occurs when the boom moves from its designed vertical position to almost a horizontal position on the water. Oil passes over or under a planing boom. Planing occurs if the tension members are poorly designed and do not hold the boom in a vertical position or if the boom is towed in currents far exceeding the critical velocity.

Structural failure—This occurs when any of the boom's components fail and the boom lets oil escape. Sometimes structural failure is so serious that the current carries the boom away. This does not happen often in normal currents and conditions. Floating debris, such as logs and ice, can contribute to structural failure.

Shallow water blockage—This type of failure occurs when rapid currents form under a boom when it is used in shallow waters. With the boom acting like a dam, the flow of water under it increases and oil is lost in several of the ways already described. Shallow water is probably the only situation in which a smaller boom might work better than a larger one. It should be noted, however, that booms are not often used in shallow water.

Tow Forces

Another consideration for the use of oil spill booms is the towing force required to pull a boom. This can be quite considerable, especially when there are waves and a strong current opposed to the pull. Figure 6.6 illustrates the towing forces in two typical situations. The forces exerted by waves and current can easily destroy a boom, tear a towline, or cause a failure of towing by a smaller tow vessel.

Ancillary Equipment

A wide variety of ancillary equipment is used with booms. *Handholds* are often installed on smaller booms that can be lifted by hand and *lifting points* are installed on larger booms for lifting by crane. Without such provision for lifting, booms must often be lifted using ropes or cables placed around the boom, which can cause damage.

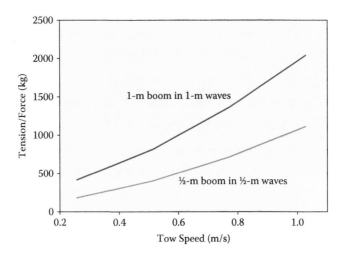

FIGURE 6.6
Towing forces calculated for a 60-meter (200-foot) boom.

All booms have some form of *end connector* for joining them to other booms or to other pieces of hardware for towing or anchoring. Although there are some standard connectors, they also vary among different manufacturers of booms, which can complicate the hookup.

Towing bridles and *towing paravanes* are pieces of equipment designed to be attached to the boom so that it can be towed without being submerged or stressed. Booms are usually towed to the site of a spill in a straight line and must withstand stresses associated with this mode of transport. Anchors, anchor attachments, and lines are also available for use with booms.

Booms are often stored on reels or in special containers designed for fast and efficient deployment. This is particularly important with heavier fire-resistant booms, as a 50-m section of such a boom could weigh hundreds of kilograms.

Sorbent Booms and Barriers

Sorbent booms are specialized containment and recovery devices made of porous sorbent material such as woven or fabric polypropylene, which absorbs the oil while it is being contained. Sorbent booms are used when the oil slick is relatively thin for the final "polishing" of an oil spill, to remove small traces of oil or sheen, or as a backup to other booms. Sorbent booms are often placed off a shoreline that is relatively unoiled or freshly cleaned to remove traces of oil that may recontaminate the shoreline. They are not

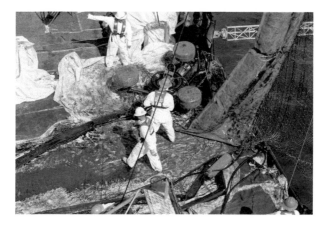

PHOTO 6.4
A boom that has been used, retains a large amount of oil and must be decontaminated. Here, a boom is being pulled onto a ship for transfer to a decontamination station. (Photo from the U.S. Coast Guard Web site: http://cgvi.uscg.mil.)

PHOTO 6.5
A boom is used in a diversionary system to collect oil. This collection was quite successful. (Photo courtesy of Applied Fabric Technologies, Inc.)

absorbent enough to be used as a primary countermeasure technique for any significant amount of oil.

Sorbent booms require considerable additional support to prevent breakage under the force of strong water currents. They also require some form of flotation so they will not sink once saturated with oil and water. Oil sorbent booms must also be removed from the water carefully to ensure that oil is not forced from them and the area recontaminated.

PHOTO 6.6
A shoreline seal boom (foreground) deployed to protect a sensitive area. (Photo from the U.S. Coast Guard Web site: http://cgvi.uscg.mil.)

PHOTO 6.7
A U configuration that is leaking a small amount of oil behind due to entrainment. (Photo from the U.S. Coast Guard Web site: http://cgvi.uscg.mil.)

PHOTO 6.8
Booms being deployed in a river. Note the shallow angle that the booms in the center of the river have been placed to avoid failure. (Photo courtesy of Applied Fabric Technologies, Inc.)

PHOTO 6.9
A boom being used on a calm lake to successfully contain oil near shore. (Photo courtesy of Environment Canada.)

PHOTO 6.10
A sorbent boom is used to attempt to contain a diesel oil spill under the snow. (Photo courtesy of Environment Canada.)

Special Purpose Booms

A variety of special purpose booms is available. A *tidal seal boom* floats up and down, but forms a seal against the bottom during low tide. These are often used to protect beaches or other stretches of shoreline from oiling. An *ice boom* is used to contain or divert oil in ice-infested waters. It is not used to contain or divert ice. An ice boom usually has slots at the waterline so that oil and water can pass through but ice cannot. A **bubble barrier** consists of an underwater air delivery system, which creates a curtain of rising bubbles that deflect the oil. Bubble barriers are occasionally used at fixed facilities such as harbors and loading platforms where the water is generally calm. The concept of bubble barriers is illustrated in Figure 6.7. High-pressure **air or water streams** can also be used to contain and deflect oil. Because of their high power requirements, they are usually used only to deflect oil in front of skimmers or fixed separator systems.

Chemical barriers use chemicals that solidify the oil and prevent its spread. Large amounts of chemicals are required, however, and the potential for containment is low. *Net booms* made from fine nets are used to collect viscous oils, tar balls, and oiled debris without having the large hydrodynamic forces of a solid boom. *Oil trawls* are similar to net booms but are made in the shape of a U so that oil is contained in the net pocket.

Fire-resistant booms are used when oil is burned on site. These booms, which are made of specialized materials that withstand high heat fluxes, are discussed in Chapter 10.

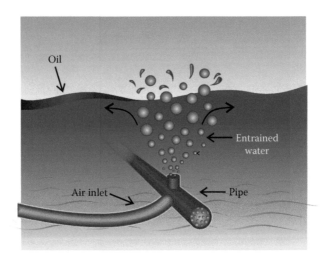

FIGURE 6.7
Bubble barrier.

PHOTO 6.11
Inflatable booms are often stored on reels for fast and convenient deployment. (Photo courtesy of Elastec/American Marine Inc.)

PHOTO 6.12
Larger booms are often stored in shipping containers or containers such as this one. (Photo courtesy of Elastec/American Marine Inc.)

7

Oil Recovery on Water

Recovery is the next step after containment in an oil spill cleanup operation. It is often the major step in removing oil from the environment. As discussed in the previous chapter, an important objective of containment is to concentrate oil into thick layers to facilitate recovery. In fact, the containment and recovery phases of an oil spill cleanup operation are often carried out at the same time. As soon as booms are deployed at the site of a spill, equipment and personnel are mobilized to take advantage of the increased oil thickness, favorable weather, and less weathered oil. After oil spreads or becomes highly weathered, recovery becomes less viable and is sometimes impossible.

This chapter covers approaches to the physical recovery of oil from the water surface, namely, skimmers, sorbents, and manual recovery. In many cases, all of these approaches are used in a spill situation. Each method has limitations, depending on the amount of oil spilled, sea and weather conditions, and the geographical location of the spill.

Alternative methods for treating oil directly on water are discussed in Chapters 9 and 10. The recovery of oil spilled on land is discussed in Chapter 12.

Skimmers

Skimmers are mechanical devices designed to remove oil from the water surface. They vary greatly in size, application, and capacity, as well as in recovery efficiency. Skimmers are classified according to the area where they are used, for example, inshore, offshore, in shallow water, or in rivers; and by the viscosity of the oil they are intended to recover, that is, heavy or light oil.

Skimmers are available in a variety of forms, including independent units built into a vessel or containment device, and units that operate in either a stationary or mobile (advancing) mode. Some skimmers have storage space for the recovered oil and some of these also have other equipment such as separators to treat the recovered oil.

The effectiveness of a skimmer is rated according to the amount of oil that it recovers as well as the amount of water picked up with the oil.

PHOTO 7.1
Booms are typically used to sufficiently thicken an oil slick for recovery. Here, a U configuration is used to thicken oil for recovery by a skimmer. (Photo from the U.S. Coast Guard Web site: http://cgvi.uscg.mil.)

Removing water from the recovered oil can be as difficult as the initial recovery. Effectiveness depends on a variety of factors including the type of oil spilled, the properties of the oil such as viscosity, the thickness of the slick, sea conditions, wind speed, ambient temperature, and the presence of ice or debris.

Most skimmers function best when the oil slick is relatively thick and most will not function efficiently on thin slicks. The oil must therefore be collected in booms before skimmers can be effectively used. The skimmer is placed wherever the oil is most concentrated in order to recover as much oil as possible. Skimmers are often placed downwind from the boom, so that the wind will push the oil toward them. Small skimmers are usually attached to light mooring lines so that they can be moved around within the slick.

Weather conditions at a spill site have a major effect on the efficiency of skimmers. All skimmers work best in calm waters. Depending on the type of skimmer, most will not work effectively in waves greater than 1 m or in currents exceeding 1 knot. Most skimmers do not operate effectively in waters with ice or debris such as branches, seaweed, and floating waste. Some skimmers have screens around the intake to prevent debris or ice from entering, conveyors or similar devices to remove or deflect debris, and cutters to deal with seaweed. Very viscous oils, tar balls, or oiled debris can clog the intake or entrance of skimmers and make it impossible to pump oil from the skimmer's recovery system.

Skimmers are also classified according to their basic operating principles: oleophilic surface skimmers; weir skimmers; suction skimmers or vacuum devices; elevating skimmers; and submersion skimmers. Each type of skimmer has distinct advantages and disadvantages, which are discussed next. Other miscellaneous devices used to recover oil are also discussed.

Oleophilic Surface Skimmers

Oleophilic surface skimmers, sometimes called **sorbent surface skimmers**, use a surface to which oil can adhere to remove the oil from the water surface. This oleophilic surface can be in the form of a disc, drum, belt, brush, or rope, which is moved through the oil on the top of the water. A wiper blade or pressure roller removes the oil and deposits it into an onboard container or the oil is directly pumped to storage facilities on a barge or on shore. The oleophilic surface itself can be steel, aluminum, fabric, or plastics such as polypropylene and polyvinyl chloride.

Oleophilic skimmers pick up very little water compared to the amount of oil recovered, which means they have a high oil-to-water recovery ratio. They can operate efficiently on relatively thin oil slicks. They are not as susceptible to ice and debris as the other types of skimmers. These skimmers are available in a range of sizes and work best with light crude oils, although their suitability for different types of oil varies with the design of the skimmer and the type of oleophilic surface used. The *disc skimmer* is a common type of oleophilic surface device. A schematic of a disc skimmer is shown in Figure 7.1. The discs are usually made of either polyvinyl chloride or steel. Disc skimmers work best with light crude oil and are well suited to working in small waves and among weeds or debris. These skimmers are usually small and can be deployed by one or two people. Disadvantages are that the recovery rate is slow and they work poorly with light fuels or heavy oils. Some new disc skimmers are grooved so they present more surface area to the oil and this can result in improved recovery.

The *drum skimmer* is another type of oleophilic surface skimmer. A schematic of a drum skimmer is shown in Figure 7.2. The drums are made of either a proprietary polymer or steel. The drum skimmer works relatively well with fuels and light crude, but is ineffective with heavy oils. New innovations to this skimmer include grooves in the drum to increase surface area. Drum skimmers are often smaller in size like the disc skimmer.

Belt skimmers are constructed of a variety of oleophilic materials ranging from fabric to conveyor belting. An adsorbent belt skimmer is shown in Figure 7.3. Most belt skimmers function by lifting oil up from the water surface to a recovery well. The motion of the belt through the water drives oil away from the skimmer, however, oil must be forced to the belt manually or with a water spray. Some belt skimmers have been designed to overcome this problem, including one that pumps the oily water through a porous belt and the *inverted belt skimmer*, which carries the oil under the water. The oil is subsequently removed from the belt by scrapers and rollers after the belt

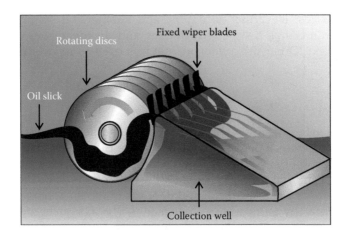

FIGURE 7.1
Schematic of a disc skimmer.

PHOTO 7.2
A disc skimmer working in ice and between a dock and a ship. (Photo courtesy of Environment Canada.)

returns to a selected position at the bottom of the skimmer. Belt skimmers of all types work best with heavier crudes and some are specially constructed to recover tar balls and very heavy oils. Belt skimmers are large and are usually built into a specialized cleanup vessel.

Brush skimmers use tufts of plastic attached to drums, chains, or belts to recover the oil from the water surface. Figure 7.4 shows a brush drum skimmer and Figure 7.5 shows a brush belt or chain skimmer. The oil is usually removed from the brushes by wedge-shaped scrapers. Brush skimmers are

PHOTO 7.3
A grooved disc skimmer under test. (Photo courtesy of Elastec/American Marine Inc.)

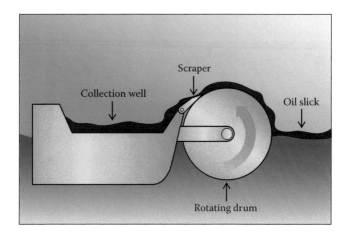

FIGURE 7.2
Schematic of a drum skimmer.

particularly useful for recovering heavier oils, but are ineffective for fuels and light crudes. Some skimmers include a drum for recovering light fuels and a brush for use with heavier oils. These skimmers can also be used with limited amounts of debris or ice. Brush skimmers are available in a variety of sizes, from small portable units to large units installed on specialized vessels or barges.

Rope skimmers remove oil from the water surface with an oleophilic rope of a polymer material, usually polypropylene. A schematic of

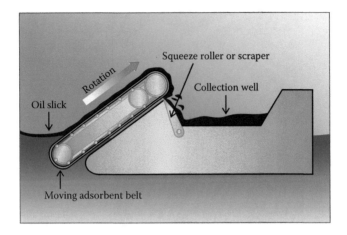

FIGURE 7.3
Schematic of an adsorbent belt skimmer.

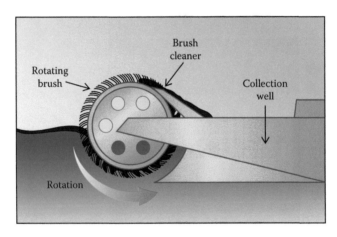

FIGURE 7.4
Schematic of a brush drum skimmer.

a rope skimmer, sometimes also called a rope mop skimmer, is shown in Figure 7.6. Some skimmers have one or two long ropes that are held in the slick by a floating, anchored pulley. Others use a series of small ropes that hang down to the water surface from a suspended skimmer body. The rope skimmer works best with medium viscosity oils and is particularly useful for recovering oil from debris- and ice-laden waters. Rope skimmers vary in size from small portable units to large units installed on specialized vessels or barges.

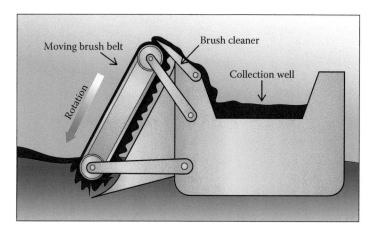

FIGURE 7.5
Schematic of a brush belt skimmer.

PHOTO 7.4
A large brush drum skimmer being lowered from its deployment module. (Photo courtesy of
Aqua-Guard Spill Response Inc.)

PHOTO 7.5
A brush chain skimmer collecting oil from wood debris. (Photo courtesy of Lamor Corporation.)

PHOTO 7.6
A brush drum skimmer recovering heavy oil. (Photo courtesy of Lamor Corporation.)

PHOTO 7.7
A brush chain skimmer being used to recover oil from ice. (Photo courtesy of Lamor Corporation.)

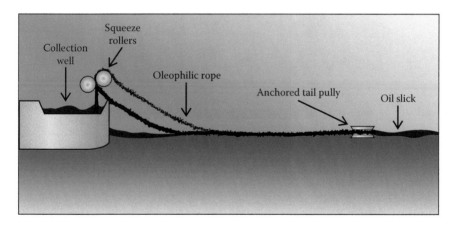

FIGURE 7.6
Schematic of a rope skimmer.

Weir Skimmers

Weir skimmers are a major group of skimmers that use gravity to drain the oil from the surface of the water into a submerged holding tank. The configuration of a weir skimmer is illustrated in Figure 7.7. In their simplest form, these devices consist of a weir or dam, a holding tank, and a connection to an external or internal pump to remove the oil. Many different models and sizes of weir skimmers are available.

One problem with some weir skimmers is their tendency to rock back and forth in choppy water, alternately sucking in air above the slick and water below. This increases the amount of water and reduces the amount of oil recovered. Some models include features for self-leveling and adjustable

PHOTO 7.8
A weir skimmer recovering emulsified oil. The water spray is being used to drive the oil toward the skimmer, which is in the apex of the boom. (Photo from the U.S. Coast Guard Web site: http://cgvi.uscg.mil.)

skimming depths so that the edge of the weir is precisely at the oil–water interface, minimizing the amount of water collected. Many new designs feature weir skimmers at the center of three or four floats, which keeps the weir edge at the water–oil interface.

Weir skimmers do not work well in ice and debris or in rough waters, and they are not effective for very heavy oils or tar balls. Weir skimmers are economical, however, and they can have large capacities. They are best used in calm, protected waters.

Suction or Vacuum Skimmers

Suction or vacuum skimmers use a vacuum to remove oil from the water surface. Often the "skimmer" is only a small floating head connected to an external source of vacuum, such as a vacuum truck. The head of the skimmer is simply an enlargement of the end of a suction hose and a float. The principle of operation of one type of suction skimmer is shown in Figure 7.8.

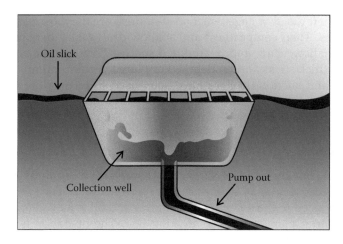

FIGURE 7.7
Schematic of one type of weir skimmer.

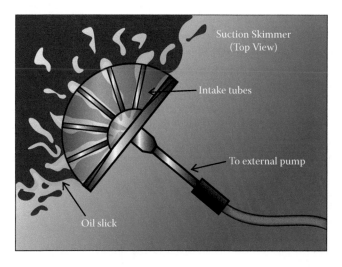

FIGURE 7.8
Schematic of a suction skimmer.

Suction skimmers are similar to weir skimmers in that they sit on the water surface, generally use an external vacuum pump system such as a vacuum truck, and are adjusted to float at the oil–water interface. They also tend to be susceptible to the same problems as weir skimmers. They are prone to clogging with debris, which can stop the oil flow and damage the pump. They also experience the problem of rocking in choppy waters, which causes massive water intake, followed by air intake. Their use is restricted to light to medium oils.

Despite their disadvantages, suction skimmers are the most economical of all skimmers. Their compactness and shallow draft make them particularly useful in shallow water and in confined spaces. They operate best in calm water with thick slicks and no debris. Very large vacuum pumps, called air conveyors, and suction dredges have been used to recover oil, sometimes directly without a head. Both these adaptations, however, have the same limitations as smaller suction skimmers.

Elevating Skimmers

Elevating skimmers or devices use conveyors to lift oil from the water surface into a recovery area. A paddle belt or wheel or a conveyor belt with ridges is adjusted to the top of the water layer and oil is moved up the recovery device on a plate or another moving belt. The operation is similar to removing liquid from a floor with a squeegee. The oil is usually removed from the conveyer by gravity. When operating these skimmers, it is difficult to maintain the conveyor at the waterline. In addition, they cannot operate in rough waters or in waters with large pieces of debris, and cannot deal with light or very heavy oils. Elevating skimmers work best with medium to somewhat heavy oils in calm waters. They are generally large and are sometimes built into specialized vessels.

Submersion Skimmers

Submersion skimmers use a belt or inclined plane to force the water beneath the surface as shown in Figure 7.9. The belt or plane forces the oil downward toward a collection well where it is removed from the belt by a scraper or by gravity. The oil then flows upward into the collection well, and is removed

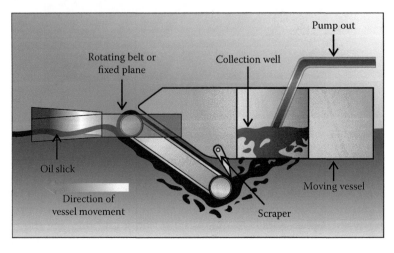

FIGURE 7.9
Schematic of a submersion skimmer.

by a pump. Submersion skimmers move faster than other skimmers and can therefore cover a large area, making them suitable for use at larger spills. They are most effective with light oils with a low viscosity and when the slick is relatively thin. Disadvantages include a poor tolerance to debris compared to other skimmers and they cannot be used in shallow waters. Submersion skimmers are larger than other types of skimmers and are usually mounted on a powered vessel.

Other Devices

Many other devices are used to recover oil. Several skimmers combine some of the principles of operation already discussed. For example, one skimmer uses an inverted belt both as an oleophilic skimmer and a submersion skimmer. A fish trawl has been modified by adding an oil outlet. Once the trawl is filled, however, usually with water, it is almost impossible to maintain a dynamic balance in the trawl so that further oil can enter. Regular fishing nets and fishing boats have been used to recover extremely large tar balls, but the oil fouls the nets, making disposal or expensive cleanup necessary. Garbage-collecting vessels have been successfully used to remove oiled debris or tar balls.

Skimmer Performance

A skimmer's performance is affected by a number of factors including the thickness of the oil being recovered, the extent of weathering and emulsification of the oil, the presence of debris, and weather conditions at the time of recovery operations.

A skimmer's overall performance is usually determined by a combination of its recovery rate and the percentage of oil recovered. The recovery rate is the volume of oil recovered under specific conditions. It is measured as volume per unit of time (e.g., cubic meter per hour [m^3/h]) and is usually given as a range. If a skimmer takes in a lot of water, it is detrimental to the overall efficiency of an oil spill recovery operation. The results of recent performance testing on various types of skimmers are given in Table 7.1.

In addition to these characteristics, other important measures of a skimmer's performance include the amount of emulsification caused by the skimmer, its ability to deal with debris, ease of deployment, ruggedness, applicability to specific situations, and reliability.

Special-Purpose Ships

Special-purpose ships have been built specifically to deal with oil spills. Some ships have been built with a hull that splits to form a V-shaped containment boom with skimmers built into the hull, although this requires very expensive design features so the ship can withstand severe weather conditions. Other ships have been built with holes in the hull to hold skimmers, with

TABLE 7.1

Performance of Typical Skimmers

Skimmer Type	Recovery Rate (m³/hr) for Given Oil Type[a]				Percent Oil[b]
	Diesel	Light Crude	Heavy Crude	Bunker C	
Oleophilic Skimmers					
Small disc	0.4 to 1	0.2 to 2			80 to 95
Large disc		10 to 20	10 to 50		80 to 95
Brush drum	0.2 to 0.8	0.5 to 20	0.5 to 2	0.5 to 2	80 to 95
Brush belt	0.4 to 1	15 to 30	1 to 10	1 to 10	80 to 95
Large drum		10 to 30			80 to 95
Small drum	0.5 to 5	0.5 to 5			80 to 95
Large belt	1 to 5	1 to 20	3 to 20	3 to 10	75 to 95
Inverted belt		10 to 30			85 to 95
Rope mop		2 to 20	2 to 10		80 to 95
Weir Skimmers					
Small weir	0.2 to 10	0.5 to 5	2 to 20		20 to 80
Large weir		30 to 100	5 to 10	3 to 5	50 to 90
Advancing weir	1 to 10	5 to 30	5 to 25		30 to 70
Elevating Skimmers					
Paddle conveyer		1 to 10	1 to 20	1 to 5	10 to 40
Submersion Skimmers					
Large	0.5 to 1	1 to 80	1 to 20		70 to 95
Suction Skimmers					
Small	0.3 to 1	0.3 to 2			3 to 10
Large trawl unit		2 to 40			20 to 90
Large vacuum unit		3 to 20	3 to 10		10 to 80

[a] Recovery rate depends on the thickness of the oil, type of oil, sea state, and many other factors.

[b] This is the percentage of oil in the recovered product. The higher the value, the less the amount of water, and thus the better the recovered oil.

sweeps mounted on the side to direct oil to the skimmer area. Many small vessels have been custom-built to hold skimmers.

Sorbents

Sorbents are materials that recover oil through either absorption or adsorption. They play an important role in oil spill cleanup and are used in the following ways: to clean up the final traces of oil spills on water or land;

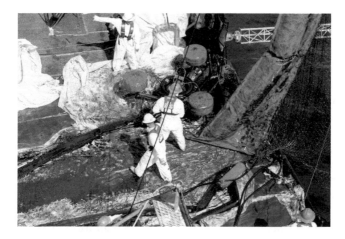

PHOTO 7.9
Sorbent sheeting used to keep oil from messing a ship's deck during the recovery of a boom and skimmer. (Photo from the U.S. Coast Guard Web site: http://cgvi.uscg.mil.)

PHOTO 7.10
Sorbent boom being placed in the water in the event that some light oil enters the area behind the regular boom. (Photo from the U.S. Coast Guard Web site: http://cgvi.uscg.mil.)

as a backup to other containment means, such as sorbent booms; as a primary recovery means for very small spills; and as a passive means of cleanup. An example of such passive cleanup is when sorbent booms are anchored off lightly oiled shorelines to absorb any remaining oil released from the shore and prevent further contamination or reoiling of the shoreline.

Sorbents can be natural or synthetic materials. Natural sorbents are divided into organic materials, such as peat moss or wood products, and inorganic materials, such as vermiculite or clay. Sorbents are available in a loose form, which includes granules, powder, chunks, and cubes, often contained in bags, nets, or "socks." Sorbents are also available formed into pads, rolls, blankets, and pillows. Formed sorbents are also made into sorbent booms and sweeps. One type of plastic sorbent is formed into flat strips or "pom-poms," which are useful for recovering very heavy oils.

The use of synthetic sorbents in oil spill recovery has increased in the last few years. These sorbents are often used to wipe other oil spill recovery equipment, such as skimmers and booms, after a spill cleanup operation. Sheets or rolls of sorbent are often used for this purpose. Synthetic sorbents can often be reused by squeezing the oil out of them, although extracting small amounts of oil from sorbents is sometimes more expensive than using new ones. Furthermore, oil-soaked sorbent is difficult to handle and can result in minor releases of oil between the regeneration area and the area where the sorbent is used.

The capacity of a sorbent depends on the amount of surface area to which the oil can adhere as well as the type of surface. A fine porous sorbent with many small capillaries has a large amount of surface area and is best for recovering light crude oils or fuels. Sorbents with a coarse surface would be used for cleaning up a heavy crude oil or Bunker. Pom-poms intended for recovering heavy Bunker or residual oil consist of ribbons of plastic with no capillary structure. General purpose sorbents are available that have both fine and coarse structure, but these are not as efficient as products designed for specific oils.

Some sorbents are treated with oleophilic (oil-attracting) and hydrophobic (water-repelling) agents to improve the ability of the material to preferentially absorb oil rather than water. As natural sorbents often recover large amounts of water along with the oil, they can be treated to inhibit water uptake. This type of treatment also increases the ability of certain sorbents to remain afloat.

The performance of sorbents is measured in terms of total oil recovery and water pickup, similar to skimmers. "Oil recovery" is the weight of a particular oil recovered compared to the original weight of the sorbent. For example, highly efficient synthetic sorbent may recover up to 30 times its own weight in oil and an inorganic sorbent may recover only twice its weight in oil. The amount of water picked up is also important, with an ideal sorbent not recovering any water. Some results of performance testing of typical sorbents with various types of oils are given in Table 7.2.

There are a number of precautions that must be considered when using sorbents. First, the excessive use of sorbents at a spill scene, especially in a granular or particulate form, can compound cleanup problems and make it impossible to use most mechanical skimmers. Sorbents may cause plugging in discharge lines or even in the pumps themselves. Second, sorbents that sink should not be used, as they could be harmful to the environment. Sinking is a problem with many sorbents such as untreated peat moss,

TABLE 7.2

Performance of Some Sorbents

Sorbent Type	Typical Oil Recovery with Oil Type (weight:weight)[a]				Percent Oil[b]
	Diesel	Light Crude	Heavy Crude	Bunker C	
Synthetic Sorbents					
Granular synthetic	2	2	1		90+
Polyester pads	7	9	12	20	90+
Polyethylene pads	25	30	35	40	90+
Polyolefin pom-poms	2	2	3	8	90+
Polypropylene pads	6	8	10	13	90+
Polypropylene pom-poms	3	6	6	15	90+
Polyurethane pads	20	30	40	45	90+
Natural Organic Sorbents					
Bark or wood fiber	1	3	3	5	70
Bird feathers	1	3	3	2	80+
Cellulose (treated) squares	3	4	6		70
Corn cobs	1	1	1		70
Cotton pads	10	15	20		70
Hair	2	6	6		70
Peat moss	2	3	4	5	80+
Treated peat moss	5	6	8	10	80+
Straw	2	2	3	4	70
Vegetable fiber	9	4	4	10	80+
Natural Inorganic Sorbents					
Clay (kitty litter)	3	3	3	2	70
Treated pearylite	8	8	8	9	70
Treated vermiculite	3	3	4	8	70
Vermulite	2	2	3	5	70

[a] Recovery depends on the thickness of the oil, type of oil, surface type, and many other factors.

[b] This is the percentage of oil in the recovered product. The higher the value, the less the amount of water, and thus the better the sorbent's performance.

all inorganic sorbents, and many wood products. Many countries do not allow the use of sorbents that sink in applications on water, as the oil will usually be released from the sorbent over time, and both the oil and the sorbent are very harmful to benthic life. And finally, recovery and disposal of the oiled sorbent material must be considered. As oiled sorbent is most often burned or placed in a landfill, the sorbent must retain the oil long enough so that it is not lost during recovery operations or in transport to disposal sites. The sorbent may be required to hold oil for a long time after it is in a landfill.

PHOTO 7.11
Heavy oil sorbent, pom-poms, placed along a shoreline to catch oil that may be washed ashore.
(Photo from the U.S. Coast Guard Web site: http://cgvi.uscg.mil.)

PHOTO 7.12
Peat moss being blown into an oiled marsh to sequester oil. The peat moss will not be removed.
(Photo courtesy of Elastec/American Marine Inc.)

Manual Recovery

Small oil spills or those in remote areas are sometimes recovered by hand. Heavier oils are easier to remove this way than lighter oils. Spills on water close to shorelines are sometimes cleaned up with shovels, rakes, or by cutting the oiled vegetation. Hand bailers, which resemble a small bucket on the end of a handle, are sometimes used to recover oil from the water surface. Manual recovery is tedious and may involve dangers, such as physical injury from falls on the shore. Most shoreline cleanup is usually done manually as discussed in Chapter 11.

8

Separation, Pumping, Decontamination, and Disposal

After oil is recovered from the water surface or from land, it must be temporarily stored, the water and debris separated from it, and the oil recycled or disposed. Pumps are used to move the oil from one process to another. This chapter covers temporary storage, separation, equipment decontamination, and disposal, as well as the types of pumps used for oil transfer. Storage, separation, and disposal are all crucial parts of a cleanup operation. In many cleanups, recovery has actually stopped because there was no place to put the recovered oil.

Temporary Storage

When oil is recovered, sufficient storage space must be available for the recovered product. The recovered oil often contains large amounts of water and debris, which increase the amount of storage space required.

Several types of specially built tanks are available to store recovered oil. *Flexible portable tanks*, often constructed of plastic sheeting and a frame, are the most common type of storage used for spills recovered on land, and from rivers and lakes. These are available in a range of sizes from approximately 1 to 100 m^3 and require little storage space before assembly. Most of these types of tanks do not have a roof, however, so rain or snow can enter the tank and vapors can escape. *Rigid tanks*, which are usually constructed of metal, are also available and are often used at sea.

Pillow tanks, constructed of polymers and heavy fabrics, are usually used to store oil recovered on land. These are placed on a solid platform so that rocks cannot puncture the tank when full. Pillow tanks are also sometimes used on the decks of barges and ships to hold oil recovered at sea. Oil recovered on land is often stored in stationary tanks built for other purposes, and in dump trucks and modular containers lined with plastic. Recovered oil can also be temporarily stored in pits or berms lined with polymer sheets, although this open type of storage is not suitable for volatile oils.

Towable, flexible tanks, usually bullet shaped, are also used to contain oil recovered at sea. Their capacity varies but they can hold up to several tons.

PHOTO 8.1
A flexible tank being towed at sea. This temporary storage means is useful for lighter oils.
(Photo courtesy of Elastec/American Marine Inc.)

PHOTO 8.2
A portable flexible tank used to store recovered Bunker C oil. (Photo courtesy of Environment
Canada.)

These tanks are also constructed of polymers, with fabric materials sometimes used as a base. Since most oils are less dense than water, these tanks
will float throughout the recovery process. When full, these tanks can be
difficult to maneuver, however, and they can be difficult to empty, especially
if the oil is viscous and contains debris.

Oil recovered at sea is often temporarily stored in *barges*. Many cleanup
organizations have barges that are used solely for storing recovered oil and

PHOTO 8.3
Heavy recovered oil is stored temporarily in a steel bin. (Photo courtesy of the U.S. National Oceanic and Atmospheric Administration [NOAA].)

PHOTO 8.4
Two steel storage tanks on the deck of a ship. This is a common way of storing recovered oil. (Photo from the U.S. Coast Guard Web site: http://cgvi.uscg.mil.)

lease barges for use at larger spills. Recovered oil is also stored in the holds of ships. This is more economical than using designated tanks on land, especially when the recovered oil has to be stored for long periods of time until a final disposal method is found. Drums, small tanks, livestock watering troughs, and even bags have also been used to contain oil from smaller spills, both on land and at sea.

PHOTO 8.5
A tank truck on top of a barge. The tank truck is being used as a separation vessel and the barge is being used to store the recovered oil. (Photo from the U.S. Coast Guard Web site: http://cgvi.uscg.mil.)

Pumps

Pumps play an important role in oil spill recovery. They are an integral part of most skimmers and are also used to transfer oil from skimmers to storage tanks. Pumps used for recovered oil differ from water pumps in that they must be capable of pumping very viscous oils and dealing with water, air, and debris. The three basic types of pumps used for pumping oil recovered from spills are centrifugal pumps, vacuum systems, and positive displacement pumps. The operating principles of some pumps are shown in Figure 8.1.

Centrifugal Pumps

Centrifugal pumps have a spinning vane that moves the liquid out of a chamber by centrifugal force. These pumps, which are regularly used for pumping water and wastewater, are not designed for use with oil and are generally not capable of dealing with material more viscous than light crudes. They are economical and universally available.

Vacuum Systems

Vacuum systems consist of vacuum pumps and tanks mounted on a skid or truck. The vacuum pump creates a vacuum in the tank and the oil moves directly through a hose or pipe to the tank from the skimmer or the source

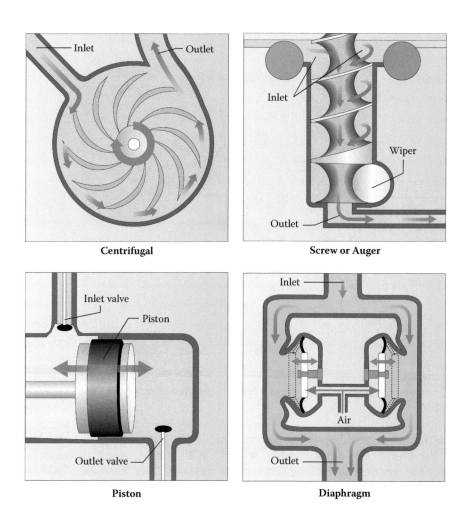

FIGURE 8.1
The operating principles of different types of pumps.

of the oil. The oil does not go through the pump but moves directly from its source into the tank.

Vacuum systems can handle debris, viscous oils, and the intake of air or water. The vacuum tank requires emptying, however, which is usually done by opening the entire end of the tank and letting the material move out by gravity.

Positive Displacement Pumps

Positive displacement pumps are often built directly into skimmers to recover more viscous oils. These pumps have a variety of operating principles, all of which have some common schemes. Oil enters a chamber in

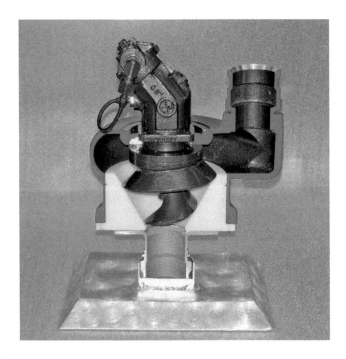

PHOTO 8.6
A cutaway of a positive displacement screw pump. (Photo courtesy of Elastec/American Marine Inc.)

the pump where it is pushed by a moving blade, shoe, or piston to the exit of the pump. The oil and other material with it must move through the chamber because there is no alternative passage, thus the name "positive displacement."

The *screw* or *auger pump* is a common type of positive displacement pump. The oil drops into part of the screw and is carried to an output. Wiper blades remove oil from the auger flights to prevent it from remaining on the auger shaft. The screw pump can deal with very viscous oils and is often built into skimmers. A *gear* or *lobe pump* uses gears or lobes mounted on a shaft to accomplish the positive displacement of oil through a chamber. Neither of these pumps can handle debris or highly viscous oil.

The *diaphragm pump* uses a flexible plate or diaphragm to move oil from a chamber. This type of pump usually requires a valve, which limits its use to material that can pass through the valve making it unsuitable for oil containing debris.

A *vane pump*, which uses a movable metal or polymer plate to move oil in a chamber, functions in a manner similar to a centrifugal pump but with positive displacement. The *peristaltic pump* uses a hose that is progressively squeezed by rollers moving along the top of its surface. As the oil never comes into contact with any material other than the hose, this pump is suitable for use with a variety of hazardous materials. Both the vane pump and peristaltic pump can handle medium viscosity oils and small debris.

A piston-like plunger in the *sliding shoe pump* moves oil along between the input and output ports. This pump does not require valves, although certain models do include them. The *piston pump* is similar to the sliding shoe pump except that oil is simply pushed out of the cylinder from the input valve to the output valve. Both sliding shoe and piston pumps can handle viscous oils but generally cannot handle debris.

Finally, the *progressive cavity pump* uses a rotating member inside a molded cylinder that together form a cavity that moves from input to output as the center is rotated. This pump can handle very viscous oils and debris, but is heavy and more expensive than other types of pumps.

Performance of Pumps

The performance of pumps is usually measured in terms of the volume displaced per unit of time at a given viscosity, suction head, and pressure head. The suction head is the maximum height that a pump can draw the target liquid, and the pressure head is the maximum height that a pump can push the target liquid. These heads are not important in most applications of pumps in oil spill recovery. It is important, however, that pumps used for pumping oil are self-priming rather than requiring that a flow of liquid be established before the pump is functional. Other important factors to consider are the pump's ability to deal with emulsions and debris, and the degree of emulsion formation that takes place in the pump itself.

Separation

As all skimmers recover some water with the oil, a device to separate oil and water is usually required. The oil must be separated from the recovery mixture for disposal, recycling, or direct reuse by a refinery. Sometimes settling tanks or gravity separators are incorporated into skimmers, but separators are more often installed on recovery ships or barges. Portable storage tanks are often used as separators, with outlets installed on the bottom of the tanks so that water that has settled to the bottom of the tank can be drained off, leaving the oil in the tank. Vacuum trucks are also used in this way to separate oil and water. Screens or other devices for removing debris are also incorporated into separators.

A *gravity separator* is the most common type of separator. In its simplest form, this consists of a large holding tank in which the oil and water mixture is held long enough for the oil to separate by gravity alone. This is referred to as the "residence time" and varies from minutes to hours. When inflow volumes are large, it can be difficult to find large enough separators to provide the long residence times required. Oil refineries have large separators that may cover several hectares, and are used for treating refinery waste and are sometimes also used to treat oil recovered from spills.

Separators are often made with baffles or other interior devices that increase the residence time and thus the degree of separation. The *parallel plate separator* is a special model of gravity separator. Many parallel plates are placed perpendicular to the flow, creating areas of low water turbulence where drops of oil can recoalesce from the water and rise to the surface.

Centrifugal separators have spinning members that drive the heavier water from the lighter oil, which collects at the center of the vessel. These separators are very efficient but have less capacity than gravity separators and cannot handle large debris. Centrifugal separators are now becoming more common. Centrifugal separators are best suited to constant amounts of oil and water. Sometimes centrifugal separators are used in tandem with gravity separators to provide a more efficient system.

As emulsions are not broken down in separators, emulsion-breaking chemicals are often added to the recovered mixture before it enters the separator. Heating the emulsions to 80°C or 90°C usually results in separation and the water can then be removed, although this process uses a lot of energy.

Separator performance is measured by the water removal efficiency and the throughput volume. Important factors affecting performance include the ability to handle small debris (larger debris is usually removed) and a wide variety of oil and water ratios, with oil content and flow rate sometimes changing suddenly.

Decontamination

Equipment and vessels used during spills often become "contaminated" or covered with oil. Before transporting this equipment further, it is decontaminated. This typically involves removal to a lined area, a high-pressure wash and treatment of the recovered water. Special areas are prepared for the decontamination of vessels, booms, or skimmers. Large vessels must, out of necessity, be decontaminated on the water, and this involves circling the vessel with booms and recovering the oil released from the vessel. Often lightly contaminated vessels are cleaned by hand using sorbent cloths. This procedure avoids the extra procedures of booming and oil recovery.

The primary tool for oil removal is high-pressure water. In the past steam was used; extra complications arise from steam including more equipment damage and heat exposure to the operators. The water released from decontamination is treated as recovered oil would be. It is separated and the oil placed in recovered-oil collection tanks.

PHOTO 8.7
Workers attach a hose to a truck at a boom decontamination station. The units in the background are oil–water separators. (Photo from the U.S. Coast Guard Web site: http://cgvi. uscg.mil.)

PHOTO 8.8
A boom being decontaminated in a semiautomatic machine. The workers in the background are packing the cleaned boom for shipment. (Photo from the U.S. Coast Guard Web site: http://cgvi.uscg.mil.)

PHOTO 8.9
A vessel being decontaminated. Note that the runoff is entirely contained. (Photo from the U.S. Coast Guard Web site: http://cgvi.uscg.mil.)

PHOTO 8.10
A worker decontaminating a hose using high-pressure water. (Photo from the U.S. Coast Guard Web site: http://cgvi.uscg.mil.)

PHOTO 8.11
Workers at a personnel decontamination station. (Photo from the U.S. Coast Guard Web site: http://cgvi.uscg.mil.)

A final note on this topic is that workers must also decontaminate their boots and clothing if covered with oil. Stations are often set up very close to embarkation points to avoid carrying contamination further.

Disposal

Disposing of the recovered oil and oiled debris is one of the most difficult aspects of an oil spill cleanup operation. Any form of disposal is subject to a complex system of local, provincial or state, and federal legislation. Unfortunately, most recovered oil consists of a wide range of contents and cannot be classified as simply liquid or solid waste. The recovered oil may contain water that is difficult to separate from the oil and many types of debris, including vegetation, sand, gravel, logs, branches, garbage, and pieces of containment booms. This debris may be too difficult to remove and thus the entire bulk material may have to be disposed.

Spilled material is sometimes *directly reused* either by reprocessing in a refinery or as a heating fuel. Some power plants and even small heating plants such as those in greenhouses can use a broad spectrum of hydro-carbon fuels. Often the equipment at refineries cannot handle oils with debris, excessive amounts of water, or other contaminants, and the cost of pretreating the oils can far exceed the value that might be obtained from using them.

Heavier oils are sometimes sufficiently free of debris to be used as *road cover* when mixed with regular asphalt. Recovered material from cleaning

PHOTO 8.12
Sand contaminated with tar balls being fed into an asphalt plant for road construction.
(Photo from the U.S. Coast Guard Web site: http://cgvi.uscg.mil.)

up beaches can be used in this way. If the material is of the right consistency, usually sand, the entire mixture might be mixed with road asphalt.

Incineration is a frequent means of disposal for recovered oil as large quantities of oil and debris can be disposed of in a relatively short time. Disadvantages are the high cost, which may include the cost of transporting the material to the facility. In addition, approval must be obtained from government regulatory authorities. Emission guidelines for incinerators may preclude simply placing material into the incinerator. Spill disposal is sometimes exempt from regulations or special permits are available. Several incinerators have been developed for disposing of either liquid or solid materials, but these all require special permits or authority to operate. In remote locations, it may be necessary to burn oiled debris directly on the recovery sites without an incinerator because it is too bulky to transport to the nearest incinerator.

Contaminated beach material is difficult to incinerate because of the sand and gravel content. There are also machines to wash oily sand or gravel. The oil recovered from this process must be separated from the wash water and then disposed of separately. Incineration should be differentiated from in-situ burning, which involves burning the material directly on site without the use of a special device. Burning debris on site is usually only applied to lightly oiled driftwood and special permission must be obtained from the appropriate authorities.

Oiled debris, beach material, and sorbents are sometimes disposed of at *landfill sites*. Legislation requires that this material not contain free oil that could migrate from the site and contaminate groundwater. Some governments have standard leachability test procedures that determine whether the material will release oil. Several *stabilization* processes have been developed to ensure that free oil does not contaminate soil or groundwater. One process uses quick lime (calcium oxide) to form a cement-like material, which can be used on roads as a dust inhibitor. *Landfarming* is the application of oil and refinery waste to land where it degrades. This practice is now banned in most jurisdictions since many oil components do not break down and contaminants, such as metals and polyaromatic hydrocarbons, are carried away from the site, often in groundwater.

Lightly contaminated water, less than 15 parts per million (ppm) by weight of oil, can usually be returned to the water body from where it came. More contaminated water may require further treatment in separators or at a municipal wastewater treatment plant.

9

Spill-Treating Agents

Treating the oil with specially prepared chemicals is another option for dealing with oil spills. An assortment of chemical spill-treating agents is available to assist in cleaning up oil. Approval must be obtained from the appropriate authorities before these chemical agents can be used. In addition, these agents are not always effective, and the treated oil may be toxic to aquatic and other wildlife.

Dispersants

Dispersant is a common term used for chemical spill-treating agents that promote the formation of small droplets of oil that "disperse" throughout the top layer of the water column. Dispersants contain surfactants, chemicals like those in soaps and detergents, that have molecules with both a water-soluble and oil-soluble component. Depending on the nature of these components, surfactants cause oil to behave in different ways in water. Surfactants or surfactant mixtures used in dispersants have approximately the same solubility in oil and water, which temporarily stabilize oil droplets in water so that the oil will disperse into the water column. This can be desirable when an oil slick is threatening a sensitive area such as a sensitive shoreline.

Two major issues associated with the use of dispersants—their effectiveness and the toxicity of the resulting oil dispersion in the water column—have generated controversy in the last 40 years. Some products used in the late 1960s and early 1970s were highly toxic and severely damaged the marine environment. Others were not effective and resulted in wasted effort. Both these issues will be discussed next.

Effectiveness of Dispersants

The effectiveness of a dispersant is determined by measuring the amount of oil that it puts into the water column then comparing it to the amount of oil that remains on the water surface. When a dispersant is working, a coffee-colored plume of dispersed oil appears in the water column and can be seen from ships and aircraft. This plume can take up to half an hour to form. If there is no such plume, it indicates little or no effectiveness. If only

PHOTO 9.1
Heavy oil after the application of a dispersant. The white plumes are dispersant plumes, the dispersant ran off the heavy oil into the sea. There is little to no dispersion here.

a white plume forms, this is the dispersant alone, also indicating little or no effectiveness.

Effectiveness is influenced by many factors, including the composition and degree of weathering of the oil, the amount and type of dispersant applied, sea energy, salinity of the water, and water temperature. The composition of the oil is the most important of these factors, followed closely by sea energy and the amount of dispersant applied. Dispersion will not occur when oil has spread to thin sheens. Below a certain thickness, the applied dispersant will interact with the water and not the oil.

As discussed in Chapter 4, some oils are prone to natural dispersion, particularly those that contain large amounts of lower molecular weight saturates. For example, diesel fuel, which contains mostly saturates, disperses well both naturally and when dispersant is added. The amount of diesel that disperses when dispersants are used compared to the amount that would disperse naturally depends primarily on the amount of dispersant entering the oil. On the other hand, oils that consist primarily of resins, asphaltenes, and larger aromatics or waxes will disperse poorly even when dispersants are applied and will in fact separate to some degree and remain on the surface. For this reason, certain products such as Bunker C are very difficult or impossible to disperse with chemical dispersants.

Laboratory studies have found that there is a trade-off between the amount (or dose) of dispersant applied and the sea energy at the time of application. In general, it was found that more dispersant is needed when the sea energy is low to yield the same amount of dispersion as when the sea energy is high. The effect of sea energy when the same amount of dispersant is used on several different types of oil is shown in Table 9.1. It can be seen that dispersants are more effective when sea energy is high than when it is low. It is also noted

TABLE 9.1

Typical Dispersant Effectiveness

Oil	Initial Dispersant Effectiveness[a]	
	At Low Sea Energy	At High Sea Energy
	(Percent of Oil in the Water Column)	
Diesel	60	95
Light crude	40	80
Medium crude	10	60
IFO 180	5	10
Bunker C	1	1

[a] The amount of dispersed oil in the water column decreases with time; in a day these amounts would decrease to about one-half and then in another day could decrease by another half.

that the amount of dispersant in the oil decreases with time and that the initial amounts of oil dispersed will be reduced to about one-half in about one day.

Effectiveness of dispersants is difficult to determine, as it is hard to accurately measure both the amount of oil in the water column and the oil remaining on the surface. Although these are easier to measure in the laboratory, testing procedures vary greatly and may not always be representative of actual conditions. When testing in the lab, important factors influencing effectiveness, such as sea energy and salinity, must be taken into consideration. It is even more difficult to measure effectiveness in the field than it is in the lab. Measurements taken in the field are best viewed as estimates, as it is difficult to take sufficient measurements at frequent enough time periods to accurately measure the concentration of oil in the water column. Accurately determining how much oil is left on the surface is a difficult task as there are no accurate methods for measuring the thickness of an oil slick and the oil in the subsurface often moves differently than the oil on the surface.

The effectiveness of dispersants is also affected by a number of factors, such as illustrated in Figure 9.1. Once the dispersant is applied, it can interact one or more of three ways upon hitting the oil on the surface. If the oil is highly viscous, the oil can run off and run into the sea, creating a white plume that is sometimes seen during dispersant applications. The dispersant may also break through the oil slick; if this occurs the oil is pushed aside or "herded." This can open wide swaths of clear water temporarily. This is not effective and the oil often returns to fill in the open swath after the dispersant enters the water. Herding occurs particularly on thin slicks or when the dispersant droplet size is of about the same dimensions as the thickness of the oil.

Ideally, the dispersant will land on and mix with the oil. The dispersant once mixed with the oil will move to the oil–water interface. Once at the interface, the effect of the dispersant will be to increase the number of small droplets formed by wave action as well as to keep these small droplets from coalescing.

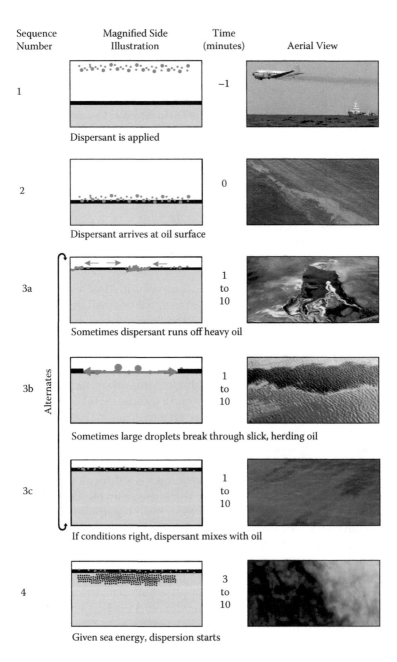

FIGURE 9.1
The effectiveness of oil dispersants as influenced by a number of factors.

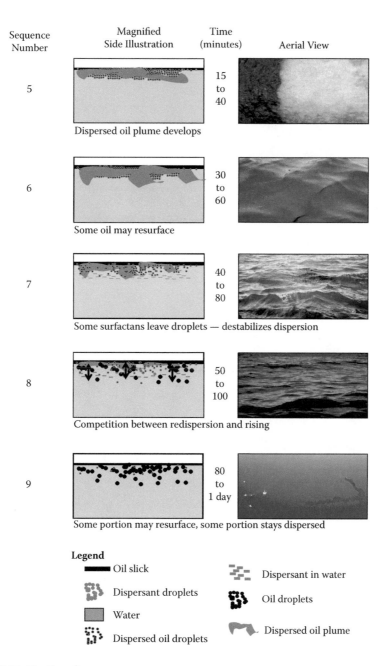

Sequence Number	Magnified Side Illustration	Time (minutes)	Aerial View
5	Dispersed oil plume develops	15 to 40	
6	Some oil may resurface	30 to 60	
7	Some surfactans leave droplets — destabilizes dispersion	40 to 80	
8	Competition between redispersion and rising	50 to 100	
9	Some portion may resurface, some portion stays dispersed	80 to 1 day	

Legend

▬ Oil slick

Dispersant droplets

Water

Dispersed oil droplets

Dispersant in water

Oil droplets

Dispersed oil plume

FIGURE 9.1 (*Continued*)

PHOTO 9.2
Coffee-colored plume indicating a good dispersion.

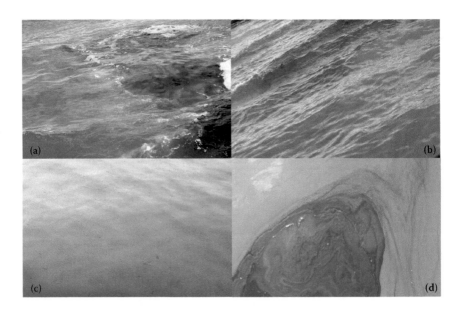

PHOTO 9.3
A collage of photos of a slick treated with dispersant in a test tank. (a) Early view of a slick after dispersant application in a test tank. (b) View of the slick about 20 minutes after dispersant application. (c) View of the slick about 90 minutes after dispersant application and after the wave generator was turned off. (d) View of the slick about 1 day after the experiment. The wave generator was off for about 22 hours.

This action is seen by the formation of a coffee-colored plume and dispersion is said to take place. Sometimes all three of these actions—runoff, herding, and effective dispersion—can take place on different parts of the same slick.

Upon forming a dispersion plume, the plume spreads. Some of the surfactant in the dispersant forming the dispersion slowly leaches into the water. This slowly destabilizes the dispersion. After about 1 to 3 hours some of the heavy components of the oil may resurface. During this time there is a competition between redispersal of the droplets by wave action and the slow, continual rise of droplets. A significant portion of the slick may resurface within about a day. As the transport of the subsurface and the surface slick often differs, the resurfacing slick may be too thin to be seen and may be in a different location than the undispersed slick.

Application of Dispersants

Dispersants are typically applied either "neat" or undiluted. Aerial spraying, which is done from small and large fixed-wing aircraft as well as from helicopters, is the most popular application method. Spray systems on small aircraft used to spray pesticides on crops can be modified to spray dispersant. Such aircraft can perform many flights in 1 day and in many different conditions. Their capacities vary from about 250 to 1000 L of dispersant. Transport aircraft with internal tanks can carry from 4000 to 12,000 L of dispersant.

Large transport aircraft, such as a Hercules, fitted with portable spray systems can carry about 20,000 L, which could treat 400,000 L of oil at a dispersant-to-oil ratio of 1:20. At a thickness of 0.5 mm, this oil would cover about 400,000 m^2 or 0.4 km^2. This treatment could be applied in as little as an hour after loading the dispersant and as many as eight flights could be flown in a day, depending on the distance from the airport to the spill.

PHOTO 9.4
A dispersant application aircraft releasing dispersant about 20 meters over a slick.

PHOTO 9.5
A large aircraft applying dispersant. (Photo courtesy of Marine Spill Response Corporation [MSRC].)

When using large aircraft, however, it can be difficult to obtain the amount of dispersant required. A co-op typically stores 100 drums or about 20,000 L of dispersant, which would be sprayed in one flyover. Further flights would have to await the arrival of more dispersant from other co-ops or production sources. An entire country's supply of dispersant can easily be consumed in 1 day if large aircraft are used.

When using helicopters, spray buckets are available in many sizes from about 500 to 2000 L. If applied at a dispersant-to-oil ratio of 1:20, 10,000 to 40,000 L of oil could be treated. If the slick were 0.5 mm thick, this would cover about 10,000 to 40,000 m^2 (or about 0.01 to 0.04 km^2). Each bucket would take about 1 to 2 hours to fill and spray over the oil. As a spill countermeasure, this rapid coverage of such a large area is appealing. Consideration of the actual effectiveness should, however, be borne in mind.

Spray systems are available for boats, varying in size from 10- to 30-m wide spray booms to tanks from 1000 to 10,000 L. New systems have as much as a 50-m wide spray with tanks on board that can carry as much as 100,000 L dispersant. As dispersant is usually diluted with seawater to maintain a proper flow through the nozzle, extra equipment is required on the vessel to control dilution and application rates. About 10,000 to 100,000 L of dispersant can be applied a day, which would cover an area of 1,000,000 m^2 or 1 km^2. As this is substantially less than could be sprayed from a single aircraft, spray boats are rarely used for a large spill. Smaller spray vessels are rarely used.

The essential elements in applying dispersant are to supply enough dispersant to a given area in droplets of the correct size and to ensure that the dispersant comes into direct contact with the oil. Droplets larger than 1000 µm will break through the oil slick and cause the oil to collect in small ribbons, which is referred to as herding. This can be detected by the rapid clearance of the oil in the dispersant drop zone without the formation of the usual coffee-colored plume in the water column. This is detrimental and wastes

PHOTO 9.6
A large aircraft applying dispersant to a noncontinuous slick. Some of the dispersant will land on the open sea. (Photo from the U.S. Coast Guard Web site: http://cgvi.uscg.mil.)

the dispersant. Herding can also occur on thinner slicks when the droplets of dispersant are smaller. The distribution of smaller droplets of dispersant is also not desirable especially when spraying from the air, as small droplets will blow away with the wind and probably not land on the intended oil slick.

Finally, it is very difficult with aerial equipment to spray enough dispersant on a given area to yield a dispersant-to-oil ratio of 1:20. The rate at which the dispersant is pumped and the resulting droplet size are critical, and a slick must often be underdosed with dispersant. Tests have shown that reapplying dispersant to the same area several times is one way of ensuring that enough dispersant is applied to the oil.

Dispersants must always be applied with a system designed specifically for the purpose. If pesticide spray equipment is used, small droplets form that may blow away and not enough dispersant is deposited onto the oil slick. Unmodified fire monitors or regular hoses from ships may not result in correct droplet sizes or quantities of dispersant per unit area. There are new nozzles designed to spray dispersants from ships. These nozzles break

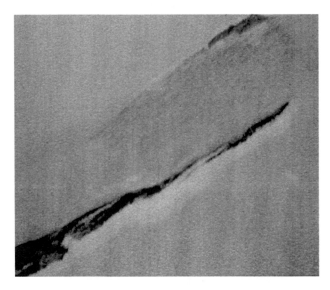

PHOTO 9.7
An image showing herding. The center of this slick was brown-black, like the edges, but disappeared temporarily after dispersant application. Within another few minutes (about 10 minutes) the slick reappears as the surfactant leaches into the water and the oil respreads.

PHOTO 9.8
A helicopter applying dispersant to an oil slick.

the spray up into droplets that are small enough to disperse the oil but large enough not to drift away in low winds.

A new method of dispersant application was tried on the *Deepwater Horizon* spill in the Gulf of Mexico. This was the direct injection of dispersants into the blowout plume near the seafloor. Further study is required before the effectiveness of this method can be determined.

PHOTO 9.9
Subsea application of dispersant during the 2010 *Deepwater Horizon* oil spill. This photo was taken 1500 meters below sea level. (Photo courtesy of the U.S. National Oceanic and Atmospheric Administration [NOAA].)

Toxicity of Dispersants

Toxicity, both of the dispersant and of the dispersed oil droplets, became an important issue in the late 1960s and early 1970s when toxic products were applied that resulted in substantial loss of sea life. Dispersants available today are less toxic (often one-hundredth as toxic) than earlier products. A standard measure of toxicity for a product is its acute toxicity (short-term dose considered on a short-time period) to a standard species such as the rainbow trout. A substance's "lethal concentration to 50% of a test population" (LC_{50}) is usually given in milligrams per liter (mg/L), which is approximately equivalent to parts per million (ppm). The specification is given with a time period, which is often 96 hours for larger test organisms such as fish. The smaller the LC_{50} number, the more toxic the product.

The toxicity of the dispersants used in the late 1960s and early 1970s ranged from about 5 to 50 mg/L measured as an LC_{50} to the rainbow trout over 96 hours. Dispersants available today vary in toxicity from 200 to 500 mg/L and contain a mixture of surfactants and a less toxic solvent. Today, some oils are more toxic than the dispersants, with the LC_{50} of diesel and light crude oil typically ranging from 20 to 50 mg/L, whether the oil is chemically or naturally dispersed. It is important to note that there are many measures of toxicity today, not just the acute ones noted earlier. There are many different toxicity measures such as long-term studies, toxicity to genetic systems, and gender modification.

The natural or chemical dispersion of oil in shallow waters can result in a greater concentration of oil in the sea that may be toxic to sea life. For example, diesel fuel spilled in a shallow bay off the Atlantic coast killed thousands of lobsters and other sea life. This occurred without the use of dispersants.

The use of dispersants remains a controversial issue and special permission is required in most jurisdictions. In some jurisdictions, their use is banned. In Canada, special permission is required from Environment Canada, through the Regional Environmental Emergencies Team (REET) or regional response team. Similarly, in the United States, special permission is required from the U.S. Environmental Protection Agency (EPA), and in waters near shore, permission is also required from the state. In both countries, products must pass standard test procedures for toxicity and effectiveness before they can be used. Only about 5 of approximately 30 proposed products are approved for use in a typical year. In summary, around the world, there is variable use of dispersants. Dispersants have been used sometimes in North America in the past 10 years and in Europe only three countries occasionally use dispersants.

The use of dispersants remains a trade-off between toxicity to aquatic life and saving birds and shoreline species. Unfortunately, dispersants are never 100% effective so both surface and aquatic life may be affected by a spill if it is treated.

Surface-Washing Agents

Surface-washing agents or beach cleaners are different from dispersants, although historically both products were sometimes referred to as dispersants. Surface-washing agents are effective in some situations, but they have not been widely accepted partially because of the confusion with dispersants. Although toxicity has been a problem with some dispersants in the past, testing has shown that the better surface-washing agents have less aquatic toxicity than dispersants and their use could prevent damage to shoreline species.

Although both products contain surfactants, those in dispersants are equally soluble in both water and oil, whereas in surface-washing agents, the surfactants are more soluble in water than in oil. Surface-washing agents operate by a different mechanism than dispersants. This mechanism is known as detergency and is similar to the use of detergents for washing clothes. In fact, dispersants and surface-washing agents may be quite different. Testing has shown that a product that is a good surface-washing agent is often a poor dispersant and vice versa.

Dispersants and surface-washing agents are used for quite different purposes. Rather than causing the oil to disperse, surface-washing agents are intended to be applied to shorelines or structures to release the oil from the surface. During low tide, the oil is sprayed with the surface-washing agent, which is then left to soak for as long as possible. It is then washed off with a low-pressure water stream in an area that has been isolated using booms.

TABLE 9.2

Effectiveness and Toxicity of Some Surface-Washing Agents

Product Description	Effectiveness of the Agent (Percentage of Oil Removed)		Toxicity (LC$_{50}$ to Rainbow Trout in 96 hours) in ppm
	In Saltwater	In Freshwater	
Approved commercial agent	55	50	>10,000
Pure d-limonene (citrus peel extract)	52	50	35
Solvent-based cleaner	44	49	25
Dispersing agent	27	25	850
D-limonene and formulation	21	23	15
Household soap	16	14	15

Skimmers are typically used to remove the released oil. Laboratory- and field-scale tests have shown that these agents substantially reduce the adhesion of the oil so that as much as 90% of the oil is released from rocks.

Environment Canada has developed a laboratory effectiveness test for surface-washing agents. This test measures the effectiveness of a product in removing weathered Bunker C from a metal trough in both saltwater and freshwater. Some typical test results are given in Table 9.2. As can be seen in the table, the most effective product, the approved commercial agent, also happens to be the least toxic. Interestingly, a natural product, d-limonene combined with a chemical, and a household cleaner are the most toxic and the least effective.

As with dispersants, the use of surface-washing agents is subject to rules and regulations in both Canada and the United States. Only a few products have passed both the effectiveness and toxicity criteria, and permission must be obtained before they can be used. Many other countries have similar legislation.

Although it has been proposed that surface-washing agents be used on land spills, this is forbidden in most jurisdictions because it moves the oil to the groundwater. It is much more difficult and expensive to clean up subsurface spills or groundwater than to physically remove a surface layer contaminated with oil.

Emulsion Breakers and Inhibitors

Emulsion breakers and inhibitors are agents used to prevent the formation of water-in-oil emulsions or to cause such emulsions to revert to oil and water. Several formulations can perform both functions. Emulsions can seriously complicate a cleanup operation by increasing the amount of material to be recovered, disposed, and stored by up to three times. Water-in-oil emulsions are so viscous that skimmers and pumps often cannot handle them.

There are different types of emulsion breakers and inhibitors, some of which are best used when little water is present, which is referred to as a closed system, and others that are best used on the open water, referred to as an open system. For example, some contain surfactants that are very soluble in water and are best used in closed systems so that they are not lost to the sea. Others contain polymers that have low water solubility and thus are best used on open water. The aquatic toxicity of the products also varies widely. The effectiveness of emulsion breakers and inhibitors is measured as the minimum dose required to break a stable emulsion or prevent one from forming.

As with dispersants, the use of emulsion breakers or inhibitors is subject to rules and regulations in Canada and the United States. Only a few agents have passed both the effectiveness and toxicity criteria, and permission is required to use them. Similar legislation exists in many countries, especially for the use of these products on open waters. In recent years, there has been very little use of these products.

Recovery Enhancers

Recovery enhancers, or viscoelastic agents, are formulations intended to improve the recovery efficiency of oil spill skimmers or suction devices by increasing the adhesiveness of oil. These agents can increase the recovery rate of sorbent surface skimmers for products like diesel fuel by up to 10 times. These products are not useful, however, with normally adhesive products like heavy crude oils and Bunker C. One recovery enhancer consists of a nontoxic polymer in the form of coiled molecular macromolecules, which increases the adhesion of one portion of the oil to the other.

Solidifiers

Solidifiers are intended to change liquid oil to a solid compound that can be collected from the water surface with nets or mechanical means. They are sometimes referred to as gelling agents or collecting agents. Collecting agents are actually a different category of agent that are the opposite of dispersants and are not yet fully developed. Solidifiers consist of cross-linking chemicals that couple two molecules or more, or polymerization catalysts that cause molecules to link to each other. Solidifiers usually consist of powders that rapidly react with and fuse the oil. Depending on the agent, about 10% to 40% by weight of the agent is required to solidify the oil, under ideal

TABLE 9.3

Effectiveness and Toxicity of Some Solidifiers

Product Description	Effectiveness of Agent (Minimum Percentage of Agent to Solidify Oil)	Toxicity (LC_{50} to Rainbow Trout in 96 hours) ppm
Cross-linking agents	10 to 40	>5000
Cross-linking agents, modified	20 to 50	>3000
Sorbent-like materials	40 to 80	>1000
Wax	100 to 250	>4000

mixing conditions. The required ratio of agent to oil for several proposed solidifiers is given in Table 9.3.

Solidifiers have not been used in the past for a number of reasons. Most important, if oil is solidified at sea, it makes recovery more difficult as skimming equipment, pumps, tanks, and separators are built to deal with liquid or very viscous liquid. Second, such a large amount of agent is required to solidify oil that it would be impossible to treat even a moderate spill. Third, the faster solidifiers react with the oil, the less likely the oil is to become solidified because the oil initially solidified forms a barrier that prevents the agent from penetrating the remaining oil. Trials at sea have shown that solidifiers often do not solidify the oil mass even when large amounts of treating agents are used.

Some vendors sell sorbents, often polymers, as solidifiers. These are not solidifiers, as they do not function by cross-linking. Rather, these products are sorbents.

Sinking Agents

Sinking agents are any material, usually minerals, that absorbs oil in water and then sinks to the bottom. Their use is banned in almost all countries, however, due to serious environmental concerns. These agents can jeopardize bottom-dwelling aquatic life and the oil is eventually released to reenter the water column in the original spill area.

Biodegradation Agents

Biodegradation agents are used primarily to accelerate the biodegradation of oil in the environment. They are used primarily on shorelines or land. They are not effective when used at sea because of the high degree of dilution and the rapid movement of oil.

Many studies have been conducted on biodegradation and the use of these agents. Hundreds of species of naturally occurring bacteria and fungi have been found that degrade certain components of oil, particularly the saturate component, which contains molecules with 12 to 20 carbon atoms. Some species will also degrade the aromatic compounds that have a lower molecular weight. Hydrocarbon-degrading organisms are abundant in areas where there is oil, such as near seeps in water or on land. Studies have shown that many of these native microorganisms, which are already thriving in the local climatic and soil conditions, are better at degrading oil than introduced species that are not yet acclimatized to local conditions.

As noted in Chapter 3, different types of oil have different potential for biodegradation, based primarily on their saturate content, which is the most degradable component. For example, diesel fuel, which is almost 95% saturates, will degrade in the right conditions. However, some types of Bunker C that contain few saturates will not degrade to any extent under any circumstances. This explains why asphalt, the asphaltene fraction, and the heavy aromatic fraction of oils are often used in building roads and in roofing shingles. These are known to be nondegradable.

Biodegradation agents include *bioenhancement agents*, which contain fertilizers or other materials to enhance the activity of hydrocarbon-degrading organisms; *bioaugmentation agents*, which contain microbes to degrade oil; and combinations of these two.

Studies have shown that adding bioenhancement agents to oil spilled on land can enhance the removal rate of the saturate and some of the aromatic fraction of the oil, so that as much as 40% of the oil is degraded in time periods from 1 month to a year. It has been found that the agents are most effective when added at an oil-to-nitrogen-to-phosphorus ratio of 100:10:1. Fertilizers that maintain the soil at a more neutral level are best for degrading oil. Fertilizers that make the soil acidic usually slow biodegradation. Fertilizers that are more oil-soluble and less water-soluble are most effective as they are not as likely to be washed away.

Bioaugmentation agents are not used as extensively as bioenhancement agents at oil-contaminated sites. This is because bioaugmentation agents add new microorganisms that are not usually as effective as stimulating existing bacteria. There are strict government regulations about introducing new, nonindigenous and possibly pathogenic species to an area. All types of biodegradation agents are subject to government regulations and approval before use.

It should be noted that although biodegradation does remove the saturates and some aromatic fractions of the oil, it can take weeks or even years to remove the degradable fraction, even under ideal conditions. Furthermore, the undegradable components of the oil, which constitute the bulk of heavier crudes, remain at the spill site, usually as a tarry mat often called "asphalt pavement." It has been found that biodegradation is useful for treating oil on grasslands or other land not used to grow crops where the undegraded asphaltenes, resins, and aromatics are not likely to pose a problem.

10

In-Situ Burning

In-situ burning is an oil spill cleanup technique that involves controlled burning of the oil at or near the spill site. The major advantage of this technique is its potential for removing large amounts of oil over an extensive area in less or about the same time than other techniques but with a distinct advantage of being a final solution. Extensive research has been conducted into in-situ burning. The technique has been used at actual spill sites for some time, especially in ice-covered waters where the oil is contained by the ice. It is now an accepted cleanup technique in several countries, whereas in others it is just becoming acceptable. During the 2010 oil spill in the Gulf of Mexico, it was used extensively and contributed greatly to the removal of oil from the surface.

The advantages and disadvantages of in-situ burning are outlined in this chapter, as well as conditions necessary for igniting and burning oil, burning efficiency and rates, and how containment is used to assist in burning the oil and to ensure that the oil burns safely. Finally, the air emissions produced by burning oil are described and the results of the many analytical studies into these emissions are summarized.

The discussion in this chapter focuses primarily on burning of oil on water. Burning of oil on shorelines and land is discussed briefly in Chapters 11 and 12.

Advantages

Burning has some advantages over other spill cleanup techniques, the most significant of which is its ability to be a final solution and its capacity to rapidly remove large amounts of oil. Burning can prevent oil from spreading to other areas and contaminating shorelines and biota.

Burning oil is a final, one-step solution. When oil is recovered mechanically, it must be transported, stored, and disposed of, which requires equipment, personnel, time, and money. Often not enough of these resources are available when large spills occur. Burning generates a small amount of burn residue, which can be recovered or further reduced through repeated burns.

PHOTO 10.1
Heavily weathered oil from the *Deepwater Horizon* spill is burned in 2010. This is one of more than 400 burns carried out at this spill. (Photo courtesy of Elastec/American Marine Inc.)

In ideal circumstances, in-situ burning requires less equipment and much less labor than other cleanup techniques. It can be applied in remote areas where other methods cannot be used because of distances and lack of infrastructure. In some circumstances, such as when oil is mixed with or on ice, it may be the only available option for dealing with an oil spill.

Finally, although the efficiency of a burn varies with a number of physical factors, removal efficiencies are generally greater than those for other response methods such as skimming and the use of chemical dispersants. During several tests and actual burns, efficiency rates as high as 98% were achieved.

Disadvantages

The most obvious disadvantage of burning oil is the large black smoke plume. The concerns revolve around toxic emissions from the large black smoke plume. These emissions are discussed in this chapter. The second disadvantage is that the oil will not ignite and burn quantitatively unless it is thick enough. Most oils spread rapidly on water and the slick quickly becomes too thin for burning to be feasible. Fire-resistant booms are used to concentrate the oil into thicker slicks so that the oil can be burned. And finally, burning oil is sometimes not viewed as an appealing alternative to collecting the oil and processing it for reuse. Reprocessing facilities for this purpose, however, are not readily accessible in most parts of the world. Another factor that discourages reuse of oil is that recovered oil often contains too many contaminants for reuse and is incinerated instead.

PHOTO 10.2
A test burn of crude oil carried out to test open water burning and measure emissions.

Ignition and What Will Burn

The first major spill incident at which burning was tried as a cleanup technique was when the *Torrey Canyon* lost oil off the coast of Great Britain in 1967. The military dropped bombs and incendiary devices on the spill, but the oil did not ignite. These results discouraged others from trying this technique. Only 2 years later, however, Dutch authorities were successful at burning test slicks both at sea and on shore. In 1970, Swedish authorities successfully burned Bunker C oil from a ship accident in ice. Since this time, many successful tests and burns have been carried out in many places. It has since been found that burning is often the only viable countermeasure for oil spills in remote locations.

Early studies of in-situ burning focused on ignition as being the key to successful burning of oil on water. It has since been found that ignition can be difficult but only under certain circumstances. Studies have shown that slick thickness is an important factor required for oil to burn and that almost any type of oil will burn on water or land if the slick can be ignited. Ignition may be difficult, however, at winds greater than 20 m/s (40 knots).

An important fact of in-situ burning is that oils can be ignited if they are at least 1- to 3-mm thick and will continue to burn down to slicks about ½- to 2-mm thick. This thickness is required to insulate the oil from the water. Sufficient heat is required to vaporize material so the fire will continue to burn. In very thin slicks, most of the heat is lost to the water and vaporization/combustion is not sustained.

PHOTO 10.3
The ignition device known as a Helitorch discharges its surplus igniter fluid into a boomed area before returning to its base.

In general, heavy oils and weathered oils take longer to ignite and require a hotter flame than lighter oils. Often a primer, such as diesel fuel, may be needed to start the combustion of heavy oils. This is also the case for oil that contains water, although oil that is completely emulsified with water may not ignite at all. Although the ignitability of emulsions with varying water concentrations is not well understood, oil containing some emulsion can be ignited and burned. Several burns have been conducted in which some emulsion or high water content in the oil did not affect either the ignitability of the oil or the efficiency of the burn. Chemical emulsion breakers can be used to break down enough of the emulsion to allow the fire to get started. As fire breaks down the water-in-oil emulsion, water content may not be a problem once the fire is actually burning.

Only limited work has been done on burning oil on shorelines. Because substrata are generally wet, optimal oil thicknesses are probably similar to those required on water, that is, from 1 to 3 mm. Oil is sometimes deposited in much thinner layers than this. Burning may cause portions of the oil to penetrate further into the sediments. Furthermore, burning oil on shorelines close to human settlements and other amenities is not desirable.

Most ignition devices burn long enough and generate enough heat to ignite most oils. Several igniters have been developed, ranging from simple devices made of plastic bottles of gelled gasoline and marine flares to sophisticated helicopter-borne devices. An interesting ignition technology is the helitorch, a helicopter-slung device which dispenses packets of burning, gelled fuel that produce a flame of 800°C lasting for up to 6 minutes. The device was developed to start backfires for the forestry industry. The helitorch is not useful for igniting heavier oils, however.

Fires at actual spills and in experiments have been ignited using much less sophisticated means. Some spills were lit using diesel-soaked paper or oil-soaked sorbent. The test burn conducted at the *Exxon Valdez* spill was ignited using a plastic bag filled with burning gelled gasoline. More than 400 burns were conducted at the Gulf of Mexico spill using homemade igniters consisting of jars of diesel fuel, marine flares, and Styrofoam floats.

It is important to understand why ignition of oil on water is difficult. Most of the heat produced by a flame on top of the oil goes up, indeed about 98% or more. Further, only the vapors in a pool fire will actually burn. The heat that goes down may not be sufficient to vaporize the oil underneath and then the fire could go out. Thus, when igniting an oil the heat must be sufficient to actually ignite a sufficient area to produce vapors for continual burning. Alternatively, it has been found that with heavy oils, a primer is needed. Testing has shown that a good practical primer is diesel fuel. The procedure is then to prime a given area, typically less than a square meter, with diesel fuel and then use an igniter. The use of a plastic bottle of diesel fuel with a railroad flare, such as the homemade devices described earlier, does both the priming and the ignition.

Burn Efficiency and Rates

Burn efficiency is measured as the percentage of starting oil removed compared to the amount of residue left. The amount of soot produced is usually ignored as it is a small amount and difficult to measure. Burn efficiency is largely a function of oil thickness. Oil thicker than about 2 to 3 mm can be ignited and will burn down to about 1 to 2 mm. If a 2-mm thick slick is ignited and burns down to 1 mm, the maximum burn efficiency is 50%. If a 20-mm thick pool of oil is ignited, however, and burns down to 1 mm, the burn efficiency is about 95%. Recent research has shown that these efficiency values are affected by other factors such as the type of oil and the amount of water content. Higher efficiency is usually achieved when towing a fire-resistant boom as the oil is continually driven to the rear, burned, and leaving only a small amount of residue unburned at the end.

Most of the residue from burning oil is unburned oil with some lighter or more volatile products removed. The residue is adhesive and therefore can be recovered manually. Residue from burning heavier oils and from very efficient burns may sometimes sink in water, although this rarely happens as the residue, when cooled, is only slightly denser than seawater.

Most oil pools burn at a rate of about 2 to 4 mm per minute, which means that the depth of oil is reduced by 2 to 4 mm a minute. Table 10.1 shows the burning characteristics of several oils. Several tests have shown that this varies with the type of oil, the degree of weathering, and the water content of the oil.

An optimal burn rate for diesel fuel and light crudes is about 5000 L of oil per m^2 per day (100 gallons per ft^2 per day). Figure 10.1 illustrates burn rates and the conversion into other units. Thus, the oil spilled from a large tanker

TABLE 10.1

Burning Properties of Various Fuels

Fuel	Burnability	Ease of Ignition	Flame Spread	Burning Rate[a] (mm/min)	Sootiness of Flame	Efficiency Range (%)
Gasoline	Very high	Very easy	Very rapid, through vapors	4	Medium	95–99
Diesel fuel	High	Easy	Moderate	3.5	Very high	90–98
Light crude	High	Easy	Moderate	3.5	High	85–98
Medium crude	Moderate	Easy	Moderate	3.5	Medium	80–95
Heavy crude	Moderate	Medium	Moderate	3	Medium	75–90
Weathered crude	Low	Difficult, add primer	Slow	2.8	Low	50–90
Crude oil with ice	Low	Difficult, add primer	Slow	2	Medium	50–90
Light fuel oil	Low	Difficult, add primer	Slow	2.5	Low	50–80
Heavy fuel oil	Very low	Difficult, add primer	Slow	2.2	Low	40–70
Lube oil	Very low	Difficult, add primer	Slow	2	Medium	40–60
Waste oil	Low	Difficult, add primer	Slow	2	Medium	30–60
Emulsified oil	Low	Difficult, add primer	Slow	1 to 2	Low	30–60

[a] Typical rates only—to convert the rate to Liter/m^2/hour multiply by 60 feet.

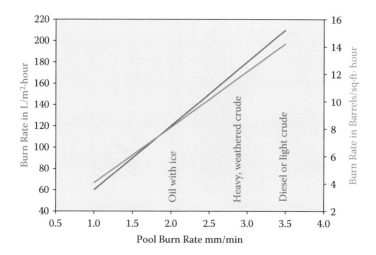

FIGURE 10.1
Burning rates for various fuels.

and covering an area about the size of the tanker's deck could be burned in about 2 days. The oil from two or three tanks from a typical tanker could be burned under the same conditions in about 6 hours. In-situ oil burning is one technique that has the potential to remove large quantities of oil in a short time.

Use of Containment

As previously discussed, oil can be burned on water without using containment booms if the slick is thick enough (2 to 3 mm) to ignite. For most crude oils, however, this thickness is only maintained for a few hours, at most, after the spill occurs. Most oil on the open sea rapidly spreads to an equilibrium thickness, which is about 0.01 to 0.1 mm for light crude oils, and about 0.05 to 0.5 mm for heavy crudes and residual oils. Such slicks are too thin to ignite and containment is required to concentrate the oil so it is thick enough to ignite and burn efficiently.

Booms are also used by spill responders to isolate the oil from the source of the spill. When considering burning as a spill cleanup technique, the integrity of the source of the spill and the possibility of further spillage is always a priority. If there is any possibility that the fire could flash back to the source of the spill, such as an oil tanker, the oil is not ignited.

Special **fire-resistant booms** are available to contain oil when using burning as a spill cleanup technique. As they must be able to withstand heat for long periods of time, these booms are constantly being tested for

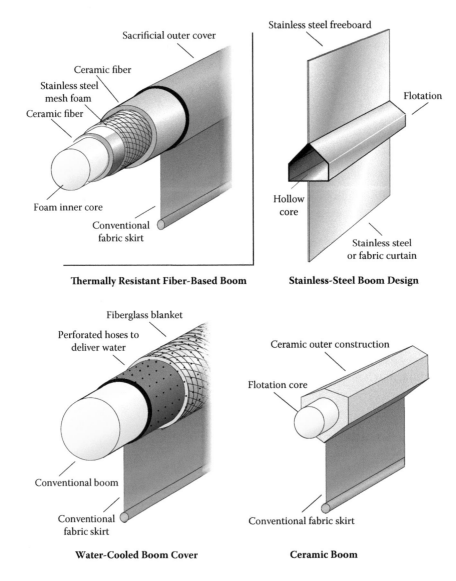

FIGURE 10.2
Fire-resistant boom designs.

fire resistance and for containment capability, and designs are modified in response to test results. Fire-resistant booms, require special handling, especially stainless-steel booms, because of their size and weight. The various designs of fire-resistant booms are shown in Figure 10.2.

Fire-resistant booms are made of a variety of materials including ceramic, stainless steel, and water-cooled fiberglass. Fire booms must withstand the high temperatures, high heat-flux as well as the mechanical forces during an oil spill burn. In addition, it is expected that a particular fire-resistant boom should withstand a multihour burn and be able to be reused several times.

One use of fire-resistant booms is to isolate the fire from the remainder of the oil. The test burn conducted at the *Exxon Valdez* site in 1989 illustrated how oil spills can be burned without threatening the source of the spill. As about four-fifths of the cargo was still in the ship, if the fire had spread, the spill could have become much larger. To avoid this risk, two fishing vessels slowly towed a fire-resistant boom on long towlines through the slick until the boom's holding capacity was reached. The oil-filled boom was then towed away from the main slick and the oil was ignited. The distance ensured that the fire could not spread back to the main slick.

During the *Deepwater Horizon* spill in the Gulf of Mexico, more than 400 burns were carried out using fire-resistant booms. In this spill, some time had passed between the rise of the oil to the surface and its movement to the zone designated for in-situ burning. Fire-resistant booms were both used to collect sufficient oil for a burn and also to isolate the burns from adjacent areas.

One approximately 200-m length of fire-resistant boom can contain about 50,000 L (11,000 gal) of oil, which, if it were a light crude, takes about 45 minutes to burn. In total, it would take about 3 hours to collect this amount of oil, tow it away from the slick, and burn it. One burn team, consisting of two tow vessels and one fire-resistant boom, could burn about three lots of oil per working shift. If there were two shifts each day, about 300,000 L of oil could be burned by each burn team in one day. A major spill could be burned even more quickly if parts of the slick could be ignited without being contained.

Oil is sometimes contained by natural barriers such as shorelines, offshore sand bars, or ice. Several successful experiments and burns of actual spills have shown that ice acts as a natural boom so that in-situ burning can be

PHOTO 10.4
The homemade igniter frequently used during the *Deepwater Horizon* spill, is lit and prepared to be put into the oil. (Photo courtesy of Elastec/American Marine Inc.)

PHOTO 10.5
A view down to an in-situ burn shows how the boom is being towed through oil and into the wind. (Photo courtesy of Elastec/American Marine Inc.)

PHOTO 10.6
A view of an in-situ burn reveals the intense fire. This is one of the burns during the *Deepwater Horizon* spill in 2010. (Photo courtesy of Elastec/American Marine Inc.)

carried out successfully for spills in ice. Oil against a shoreline can be burned if the shoreline is in a remote area and consists of cliffs, rock, gravel, or sandy slopes, and is a safe distance from any combustible material, such as forests, grass cover, or wooden structures.

Last, it should be noted that thick oils, typically greater than about 5 to 10 mm, can be burned without containment on water. Such thick oils are typically high-weathered crude oils or heavy fuel oils. Opportunities to perform uncontained burning are not frequent.

Emissions from Burning Oil

The possibility of releasing toxic emissions into the atmosphere or the water has created the biggest barrier to the widespread use and acceptance of burning oil as a spill countermeasure. Some atmospheric emissions of concern are particulate matter precipitating from the smoke plume, combustion gases, and unburned hydrocarbons. Although soot particles consist primarily of carbon particles, they also contain a number of absorbed and adsorbed chemicals. The residue left at the burn site is also a matter of concern. Possible water emissions include sinking or floating burn residue and soluble organic compounds.

Extensive studies have been recently conducted to measure and analyze all these components of emissions from oil spill burns. The emphasis in sampling has been on air emissions at ground level as these are the primary human health concern and the regulated value.

Most burns produce an abundance of particulate matter. Particulate matter at ground level is a health concern close to the fire and under the plume, although concentrations decline rapidly downwind from the fire. The greatest concern is the smaller or respirable particles, which are 2.5 µm or less in size. Concentrations at ground level (1 m) can still be above normal health concern levels (35 µm/m^3) as far downwind as 500 m from a small crude oil fire, such as from the amount of oil that could be contained in a 500-m long boom.

Polyaromatic hydrocarbons, or PAHs, are a primary concern in the emissions from burning oil, both in the soot particles and as a gaseous emission. All crude oils contain PAHs, varying from as much as 1% down to about 0.001%. Most of these PAHs are burned to fundamental gases except those left in the residue and the soot. The amount of residue left from a crude oil fire varies but generally ranges from 1% to 10%. It has been found that PAHs as gaseous emissions from oil fires are negligible. It has also been found that, compared to the original oil, the soot from several experimental burns contained a similar concentration of some PAHs of higher molecular weight and lower concentrations of PAHs of lower molecular weight. This could be a concern as the higher molecular weight PAHs are generally more toxic. This is offset, however, by the fact that in all cases the overall concentration of PAHs

PHOTO 10.7
A water-cooled boom after a burn. Some burn residue is on the boom and in the water. (Photo from the U.S. Coast Guard Web site: http://cgvi.uscg.mil.)

in the soot and residue is much less than in the original oil. These findings indicate that PAHs typically burn at the same rate as the other components of the oil and generally do not increase as a result of the fire. In summary, PAHs are not a serious concern when assessing the impact of burning oil.

The second major concern related to the emissions from burning crude oil is with the other compounds that might be produced. As this is a very broad concern, it is difficult to address. In several studies, however, soot and residue samples were extracted and "totally" analyzed in various ways. Although the studies were not conclusive, no compounds of the several hundred identified were of serious concern to human health or to the environment.

The soot analysis reveals that the bulk of the soot is carbon and that all other detectable compounds are present on this carbon matrix in quantities of parts per million or less. The compounds most frequently identified are aldehydes, ketones, esters, acetates, and acids, which are formed by incomplete oxygenation of the oil. Similar analysis of the residue shows that the same minority compounds are present at about the same levels. The bulk of the residue is unburned oil without some of the volatile compounds.

Specific analysis for the highly toxic compounds dioxins and dibenzofurans has also been carried out. These compounds were at background levels at many test fires, indicating no production by either crude or diesel fires.

Some studies have been done on the gaseous emissions from burning oil. The usual combustion products of carbon dioxide, small amounts of carbon monoxide, and sulfur dioxide, in the form of acid particulate, were found. The amount of sulfur dioxide is directly proportional to the sulfur content of the oil but is at low levels because it largely reacts with water to form the acid particulate, which in turn precipitates near the fire. Sulfur compounds in oil range from about 0.1% to 5% of the oil by weight.

TABLE 10.2

Emissions from Burning and Evaporating Slicks

Emissions	Percentage of Health Concern Levels[a] at 500 Meters			
	Burning Diesel[b]	Evaporating Diesel	Burning Light Crude	Evaporating Light Crude
Respirable particulate matter	95	0	70	0
Volatile organic compounds	5	30	10	40
PAHs on soot	0	0	5	0
Carbon monoxide	0	0	0	0
Sulfur dioxide (particulate)	0	0	2	0
Metals on soot	0	0	0.5	0
Oxygenated volatiles	1	0	0	0

[a] Health concern levels are those exposure levels that are the threshold of concern for exposure for a few hours.

[b] All estimates are based on a moderate fire of about 10,000 L burning over an area of about 50 m^2.

PHOTO 10.8
An operations crew monitors a small burn during the *Deepwater Horizon* incident. (Photo courtesy of Applied Fabric Technologies, Inc.)

When oil is burned, volatile organic compounds (VOCs) evaporate and are released. Studies have shown that benzene, toluene, xylenes, and many other volatile compounds are present in samples downwind of an oil burn. It should be noted, however, that these compounds are usually measured at higher concentrations from an evaporating oil slick that is not burning, as can be seen in Table 10.2.

Volatile oxygenated compounds are also formed when oil burns. These compounds are sometimes generally referred to as carbonyls or by their main constituents, aldehydes and ketones. Studies have shown that carbonyls from crude oil fires are at very low concentrations and are not a health concern even close to the fire. Carbonyls from diesel fires are slightly higher but are still below health concern levels.

The amount of soot produced by in-situ oil fires is not known, although estimates vary from 0.5% to 3% of the original oil volume. There are few accurate measurement techniques because the emissions from fires cover such large areas. Estimates of soot production are complicated by the fact that particulates precipitate from the smoke plume at a decreasing rate from the fire outward. When burns are conducted, heavy soot precipitation on the surface occurs near the oil pool but rapidly becomes imperceptible farther away from the burn (usually a few meters), depending on the amount of oil burned. Similarly, the smoke plume is heavy at the burn sites and dissipates within 10 to 20 km downwind.

Some concern has been expressed that the metals normally contained in oil are precipitated with soot particles. Test results from burns show that

PHOTO 10.9
In a large spill it is possible to conduct several burns at once. Here, three burns are conducted very close together. (Photo from the U.S. Coast Guard Web site: http://cgvi.uscg.mil.)

TABLE 10.3

Safe Distance Calculations[a]

Type and Area	Burn Area (Hectares [acres])	Safe Distance (Kilometers)	Safe Distance (Miles)
Crude Oil Burns			
Small area 250 m²	0.25 (0.6)	0.09	0.06
Full boom pull 500 m²	0.5 (1.2)	0.5	0.3
Large boom pull 750 m²	0.75 (1.9)	3.2	2
Diesel Burns			
Small area 250 m²	0.25 (0.6)	0.8	0.5
Full boom pull 500 m²	0.5 (1.2)	20	12.4

[a] These distances are based on calculations using actual burns and should be used as guides only. Based on PM-2.5 concentrations.

the metal concentration approaches that of emission standards very close to the fire but is negligible at about 50 m away, even when the test fire is large. It appears that much of the metal content of a crude oil fire is reprecipitated either into or very close to the fire.

Studies have enabled researchers to calculate the safe distance from burns as shown in Table 10.3. The most dangerous air emission is the smaller particulate matter, that of 2.5 μm or less. These are used to calculate the safe distances. Such distances should be used as guides only.

There has also been concern that the temperature of the water under the oil is raised when oil spills are burned on water. Measurements conducted during tests showed that the water temperature is not raised significantly, even in shallow confined test tanks. Thermal transfer to the water is limited by the insulating oil layer and is actually the mechanism by which the combustion of thin slicks is extinguished. As noted earlier, most of the heat from a fire is conducted upward and little is transferred downward.

Current thinking on burning oil as an oil spill cleanup technique is that the airborne emissions are not a serious health or environmental concern, especially at distances greater than a few kilometers from the fire. Studies have shown that emissions are low compared to other sources and generally result in concentrations of air contaminants that are below health concern levels 500 m downwind from the fire.

Summary

The use of burning as an oil spill countermeasure involves a series of trade-offs among concerns over the emissions produced, the environmental impact of a spill, the advantages of being able to remove large amounts of oil in

PHOTO 10.10
Environmental staff prepare air-sampling devices to measure the emissions from a test oil burn. (Photo courtesy of Environment Canada.)

a short period of time, and maintaining the safety of both spill workers and the source of the spill.

The potential for the use of in-situ burning must be determined based on specific conditions at the time of the spill, bearing in mind that oil is burned most easily shortly after the spill. The impact of the oil on the water and shoreline should also be considered.

In some situations, such as major spills in remote areas, burning may provide the only means of eliminating large amounts of oil quickly and safely. Burning can be used in combination with mechanical recovery and chemical dispersants. The ultimate goal is to find the right combination of equipment, personnel, and techniques to ensure that an oil spill will have the least environmental impact. In-situ burning can be a valuable tool in attaining that goal.

11

Shoreline Cleanup and Restoration

Oil spilled on water is seldom completely contained and recovered, and some of it eventually reaches the shoreline. It is more difficult and time consuming to clean shoreline areas than it is to carry out containment and recovery operations at sea. Physically removing oil from some types of shoreline can also result in more ecological and physical damage than if oil removal is left to natural processes.

The decision to initiate cleanup and restoration activities on oil-contaminated shorelines is based on careful evaluation of socioeconomic, aesthetic, and ecological factors. These include the behavior of oil in shoreline regions, the types of shoreline and their sensitivity to oil spills, the assessment process, shoreline protection measures, and recommended cleanup methods. Criteria of importance to the cleanup decision are discussed in this chapter.

Behavior of Oil on Shorelines

The fate and behavior of oil on shorelines are influenced by many factors, some of which relate to the oil itself, some to characteristics of the shoreline, and others to conditions when the oil is deposited on the shoreline, such as weather and waves. These factors include the type and amount of oil; the degree of weathering of the oil, both before it reaches the shoreline and while on the shoreline; the temperature; the state of the tide when the oil washes onshore; the type of beach substrate, that is, its material composition; the type and sensitivity of biota on the beach; and the steepness of the shore.

Other important factors are the existence of a high-tide berm on the beach, if the oil is deposited in the intertidal zone, and whether the particular length of shoreline is exposed to or sheltered from wave action. An exposed beach will often "self-clean" before a cleanup crew can perform the task, which can result in the released oil being transported to other beaches or even back to the same beach.

The extent that an oil penetrates and spreads, its adhesiveness, and how much the oil mixes with the type of material on the shoreline are all important factors in terms of cleanup. Cleanup is more difficult if the oil penetrates deeply into the shoreline. Penetration varies with the type of oil and the type of material on the shoreline. For example, oil does not penetrate much into fine beach material such as sand or clay but will penetrate extensively on a shore consisting of coarse boulders. A very light oil, such as diesel, on a cobble beach can penetrate to about a meter under some conditions and is difficult to remove. On the other hand, a weathered crude deposited on a fine-sand beach can remain on the surface indefinitely and is removed fairly easily using mechanical equipment.

The adhesiveness of the stranded oil varies with the type of oil and the degree of weathering. Most fresh oils are not highly adhesive, whereas weathered oils often are. Diesel and gasoline are relatively nonadhesive, crudes are generally moderately adhesive when fresh and more adhesive when weathered, and Bunker C is adhesive when fresh and highly adhesive when weathered. An oil that is not adhesive when it reaches the shore may get washed off, at least partially, on the next tidal cycle.

The extent of oil coverage often depends on the stage of the tide when the oil is deposited on the shoreline. At high tide, oil can be deposited above the normal tide line and often spreads over a broad intertidal area. The least amount of oiling occurs when the oil is deposited on the shoreline during the falling tide, although this is less likely to occur as the water is moving away from the shoreline. The nature of the intertidal zone, that is, its composition and slope, will often dictate the fate of the oil. If large amounts of oil are not retained in the intertidal zone, then the oil will have less impact on the area.

The fate of oil on shorelines also depends on the wave regime. Oil can be removed and carried away by energetic waves within days, whereas it can remain for decades in sheltered areas. For example, some of the oil spilled from the *Arrow* in 1970 remains in the sheltered coves of Nova Scotia to this day. Similarly, a significant amount of oil spilled from the *Metula* in 1974 remains on sheltered beaches in Chile. In both cases, the oil was Bunker C and weathering produced a crust on top of the oil. Under this crust, the oil is still relatively fresh, even after decades.

Beaches are dynamic environments that change in profile during seasonal storms. This can result in oil being buried on the beach in layers, often as deep as 1 meter, or buried oil can be brought to the surface.

Oil stranded on shorelines, especially above the high-tide line, weathers with time and becomes more adhesive, viscous, and difficult to remove. If nutrients are present and the oil is crude, limited biodegradation can take place, but this occurs slowly and only 10% to 30% of the oil is removed in 1 to 2 years. As oil stranded above the high-tide line is above the limit of normal wave action, physical removal can occur only during storm events.

Another mechanism that can significantly affect the fate of oil on shorelines is the mixing of the oil with beach material. Oil often mixes with sand and gravel on beaches and then weathers to form a hard, resilient material called "asphalt pavement," which is difficult to remove. This material may be only 1% to 30% oil by weight, which greatly increases the amount of material to be removed. Sometimes this stranded oil causes no environmental

PHOTO 11.1
A heavy layer of Bunker oil from the *Metula* spill. This oil has been in this location for more than 10 years at the time this photo was taken. Some clay material covers parts of the oil. (Photo courtesy of Ed Owens.)

concerns because the oil is entirely bound and none is lost to the water or is refloated, but there may be a concern with this oil being visible on the shoreline, depending on the location of both the oil and the shoreline.

Environmental Effects of Oil on Shorelines

Since the focus of both shoreline protection and cleanup methods is to minimize environmental damage, the environmental effects of oil on shorelines will be discussed in this section. Biota on shorelines are harmed through direct contact with the oil, ingestion of oil, smothering, and destruction of habitat and food sources. As most life on the shoreline cannot avoid the oil, its destructive effects often cannot be minimized once the oil reaches shore.

Intertidal life forms are particularly vulnerable to oil since they consist primarily of plants and animals that move slowly or not at all. It takes from months to years for an oiled intertidal zone to recolonize. Intertidal life may also be damaged by cleanup efforts, particularly by the movement of people and vehicles, and by cleaning water that is either too hot or under high pressure. A cleanup method should minimize environmental effects, not simply remove the oil at all costs. Oil should only be removed to prevent it from being refloated and oiling other shorelines. Oil stranded in the intertidal zone may cause less harm if left than if removed. If the biota is already dead, however, oil is sometimes removed so that the area can recolonize.

Oil is particularly harmful to shorebirds and mammals such as seals, sea lions, and walruses. If the beach on which they lay their eggs or give birth to pups is oiled, many of the young die after coming in contact with the oil. These areas are usually given a high cleanup priority to prevent oil from reaching the shore or to remove it quickly if it is already there.

Types of Shorelines and Their Sensitivity to Oil

The type of shoreline is crucial in determining the fate and effects of an oil spill as well as the cleanup methods to be used. In fact, the shoreline's basic structure and the size of material present are the most important factors in terms of oil spill cleanup. The structural profiles of different types of shoreline are shown in Figure 11.1.

There are many types of shorelines, all of which are classified in terms of sensitivity to oil spills and ease of cleanup. The types discussed here are bedrock, man-made solid structures, boulder beaches, pebble–cobble, mixed sand–gravel beaches, sand beaches, sand tidal flats, mud tidal flats, marshes, peat and low-lying tundra, and mangroves. These are illustrated in Figure 11.2.

Bedrock shorelines consist of rock that is largely impermeable to oil, although oil can penetrate through crevices or fractures in the rock. For this reason and because plant and animal life is scarce, bedrock shorelines are not particularly vulnerable to oil spills. Oil is more likely to be deposited in the upper tidal zone. If the shore is exposed to wave action, a significant amount of oil is likely to be removed after each tidal cycle.

Shorelines consisting of *man-made solid structures* include retaining walls, harbor walls, breakwaters, ramps, and docks, and are generally made of rocks, concrete, steel, and wood. This type of shoreline is usually considered impermeable to oil, although there are some types that are and these should be considered similar to their natural counterparts. Impermeable man-made structures are very similar to bedrock and are the least sensitive of any shoreline to oil. Recolonization by biota is usually very rapid after an oiling episode.

Boulder beaches consist primarily of materials that are more than 256 mm in diameter. These beaches are not altered by any conditions other than ice, human activity, or extreme wave conditions. Boulder beaches often give way to mud or sand tidal flats in the lower intertidal zone. Because of the large spaces between individual boulders, oil can be carried down to the sediments and remain there for years. Since animals and plants live in these spaces, oil often has a severe effect on boulder beaches. Boulder beaches are considered to be moderately sensitive to oil and do not recover rapidly from oiling.

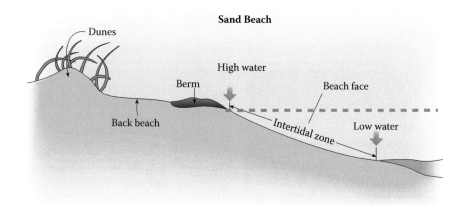

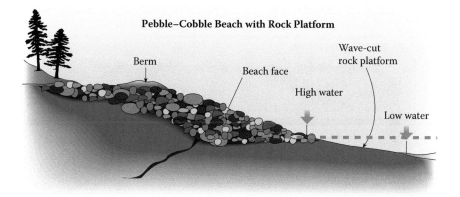

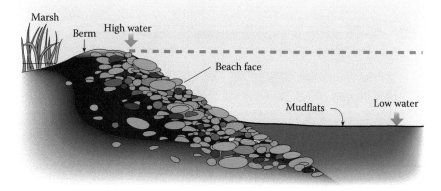

FIGURE 11.1
Shoreline profiles.

FIGURE 11.2
Photographs of some shoreline types.

PHOTO 11.2
Heavy fuel oil coats rocks on this shoreline. (Photo courtesy of the U.S. National Oceanic and Atmospheric Administration [NOAA].)

Pebble–cobble beaches consist of materials ranging in size from 2 to 256 mm. Pebbles range in size from 4 to 64 mm and cobbles from 64 to 256 mm. Some fine materials may be present in the interstitial areas between pebbles and there may also be large boulders in the area. Oil readily penetrates pebble–cobble beaches through the open spaces between the rocks. Retention of the oil may be low as it is often flushed out from the interstitial areas by natural tide or wave action. Oil will likely concentrate on the upper reaches, however, where there is little flushing action. As wave action constantly rearranges or reworks the sediments, few animals and plants are present, especially in the middle intertidal zone. Pebble–cobble beaches are not considered a sensitive beach type.

A *mixed sediment beach* consists of a variety of materials from 0.1 to 64 mm in size and possible cobbles up to 256 mm. These are sometimes called sand–gravel beaches and sometimes gravel beaches, because the larger gravel appears to predominate. Only lighter oils can penetrate sand–gravel beaches, and there is reworking of sediments, and few animals and plants. For this reason, these beaches are not considered to be particularly sensitive to oil spills. Oil from past spills can form asphalt pavement at the upper tidal reaches. Oil residence times vary but are generally shorter than on other types of beaches. As there is generally not an abundance of sand and gravel on these beaches, the profile of these beaches changes little, especially in more sheltered areas.

Sand beaches are what most people envision as a "beach." Sand is defined as a particle 0.1 to 2 mm in diameter, consisting of several different sizes and types of minerals. Coarse sand is usually defined as 0.5 to 2 mm in size, and fine sand is less than 0.5 mm. Sand beaches occur in every part of North America. On many coasts, they are often located between other types of beaches. Only lighter oils penetrate sand beaches and the residence time is likely to be short, except when oil is buried or carried to

PHOTO 11.3
Spill workers search a beach for tar balls. Recreational beaches are often used despite light oiling. (Photo from the U.S. Coast Guard Web site: http://cgvi.uscg.mil.)

PHOTO 11.4
This sandy beach was moderately oiled. (Photo courtesy of the International Tanker Owners Pollution Federation [ITOPF].)

the upper tidal areas. Oil can easily become buried in sand and over time this can result in layers of sand and oil, sometimes referred to as "chocolate layer cake." As sand beaches often do not have a high population of animals or plants, they are not considered particularly sensitive. In recreational areas, however, sand beaches are given a high cleanup priority if oiling of any type occurs.

Sand flats consist of material similar to sand beaches but are at shallow angles and never drain completely. They contain a lot of silt or very fine material. The surface layer of sand flats, which consists of a few centimeters, is dynamic and unstable. This surface layer is usually water saturated and thus impermeable to oils. Some oils may adhere to the top surface or penetrate through holes made by burrowing animals. Sand tidal flats are difficult to access and cleanup is therefore limited. Sand flats are an important bird habitat and are considered to be sensitive to oil spills.

Mudflats are similar to sand tidal flats in that they are at shallow angles and have a thin, mobile surface layer consisting of water-saturated mud that is impermeable to oil, although oil can penetrate through holes made by burrowing animals. Oil is likely to concentrate on the upper tidal zones. Mudflats are not accessible to vehicles or response personnel and thus cannot be readily cleaned. If left alone, oil is refloated and carried toward land at low tides. Like sand tidal flats, mud tidal flats are important bird habitats and are considered to be sensitive to oil spills.

Marshes are important ecological habitats that often serve as nurseries for marine and bird life in the area. Marshes range from fringing marshes, which are narrow areas beside a main water body, to wide salt marsh meadows. Salt marsh meadows often flood only during high tides in spring or during storm surges. Marshes are rich in vegetation that traps oil. Light oils can penetrate into marsh sediments through animal burrows or cracks. Heavier oils tend to remain on the surface and smother plants or animals. Oiled marshes, fresh or salt, may take years or even decades to recover. Marshes are difficult to access and entering them by foot or by vehicle can cause more damage than the oil itself. They are considered sensitive to oiling.

Peat and *low-lying tundra* are similar types of shoreline found in the Arctic regions. Although different, they have similar sensitivity and cleanup methodologies. Peat is a spongy, fibrous material formed by the incomplete decomposition of plant materials. Peat erodes from tundra cliffs and often accumulates in sheltered areas as does oil. Oil does not penetrate wet peat, but dry peat can absorb large amounts of oil. Low-lying tundra is normally dry land but is flooded by the sea at certain times of the year. Low-lying tundra includes Arctic plants and various types of sediment. Generally, oil does not penetrate tundra but it will adhere to dry vegetation on the surface. Both types of shorelines are considered moderately sensitive to oil.

Mangroves are tropical trees characterized by complex, interlaced root systems, parts of which are aerial and provide means for the trees to breathe. The term *mangrove* also refers to the complex ecosystem of which the mangrove tree is the most important component. This ecosystem can include sea grasses and many specialized organisms that are interdependent. Oil can coat the respiratory roots of mangrove trees and kill the tree within a few days. Many of the organisms in a mangrove ecosystem are sensitive. Mangrove areas are very difficult to access and to clean.

PHOTO 11.5
Seals sit atop heavily oiled rocks. This spill cost the lives of many seal pups. (Photo courtesy of Environment Canada.)

Shoreline Countermeasures Assessment Technique

Priorities for shoreline cleanup are based on a sophisticated shoreline assessment procedure. A systematic evaluation of oiled shorelines can minimize damage to the most sensitive shorelines. When an oil spill occurs, site assessment surveys are usually conducted in direct support of spill response operations. These surveys rely heavily on previously obtained data, maps, and photographs. For example, the structure of the beach is usually already mapped and recorded as part of the sensitivity mapping exercise for the area.

The following are the objectives of site assessment surveys:

- To document the oiling conditions and the physicoecological character of the oiled shoreline, using standardized procedures

PHOTO 11.6
Erosion and wave action expose bags of shoreline cleanup debris put into these sand dunes more than 20 years earlier. (Photo courtesy of the Canadian Coast Guard.)

- To identify and describe human use and effects on the shoreline's ecological and cultural resources
- To identify constraints on cleanup operations
- To verify existing information on environmental sensitivities or compare it to observations from the aerial survey
- To define the most appropriate cleanup techniques for each segment of the shoreline

A procedure for performing a site assessment survey using a Shoreline Cleanup Assessment Team (SCAT) has been documented in the literature. The SCAT concept was developed during the cleanup of the *Exxon Valdez* spill in 1989. It is a systematic and comprehensive method of data collection that can be used to evaluate shoreline oil conditions if a spill occurs and assist cleanup personnel to develop and plan response actions. The SCAT team usually includes an oil-spill geomorphologist, an ecologist, and an archeologist or a land-use specialist.

The SCAT team assesses the shoreline and assigns it a rating based on the degree of oiling, environmental resources, projected tide and wind conditions, and available transportation and other logistics. The team conducts its work according to the following three-step plan:

1. The shoreline is divided into numbered segments or an existing segmentation scheme is reviewed and adapted.

   A daily presurvey planning session is held at which decisions are made about segments to survey and required equipment and

PHOTO 11.7
Oil can be buried by wave action on sandy beaches. In this case, at least five layers of oil are
buried beneath the surface. (Photo courtesy of Environment Canada.)

supplies, existing data is reviewed, and survey members are briefed
on specific plans for the survey.

Shoreline segments are divided into zones of supratidal, upper
tidal, mid-intertidal, and lower intertidal zones. See Figure 11.1 for
illustrations of these zones.

2. The affected area is surveyed by ground surveys, aerial reconnaissance, or aerial videos. Usually all three methods are combined. The
   team completes observations and measurements on a segment, produces a sketch map of the site, and fills in forms or checklists of
   observations on the site. During the fieldwork, the SCAT team docu-
   ments the distribution and character of stranded oil, the amount and
   location of subsurface oil, shoreline characteristics, and the character
   of the substrate. Ecological and human resources in the segment are
   documented. Currently, much of this information is recorded using
   field computers or tablets.

3. After the fieldwork is finished, all forms, maps, field notes, and photo logs are completed and submitted, along with a daily report, to the command center. These are then synthesized into summary reports. Spill cleanup action is then carried out on the basis of these surveys.

PHOTO 11.8
Lightly oiled sediment is transferred to the surf line for further cleaning. This is known as surf washing.

PHOTO 11.9
Workers clear an oiled reedbed and shoreline by hand. (Photo courtesy of Environment Canada.)

Shoreline Protection Measures

Shoreline protection measures are response activities carried out on or near the shoreline, rather than on open water, to prevent the shoreline from becoming oiled or to protect vulnerable shore resources such as plants and wildlife. The objectives of shoreline protection measures are to prevent oil from contacting or collecting on certain shorelines, to minimize effects of the oil, to avoid causing more damage than the oil would by itself, to minimize waste, and to use cleanup resources in a safe, effective manner.

The use of **containment booms** to deflect the oil away from sensitive shoreline is the most common measure. If possible, oil is deflected to dock areas or loading ramps where it is easier to recover. Oil can be deflected away from sensitive shoreline to less sensitive shoreline, which may be easier to clean. For example, oil could be deflected away from a marsh to a sandy section of shoreline, where cleanup would be easier and where the oil causes less environmental damage. Booms can also be used to surround and protect a sensitive area from the oil. Booms can only be used, however, if the direct current does not exceed 0.5 m/s (about 1 knot) and the waves do not exceed the boom capabilities. Chapter 6 covers several topics on the use of containment booms.

Intertidal or *shoreline boom* is a special type of boom with a water-filled chamber in the lower section. This chamber creates a seal between the shoreline and the water so that oil cannot reach the shoreline. This boom can be used in some of the same ways as regular booms, but is best used to form a barrier perpendicular to and directly on the shore. Regular booms anchored to the shoreline are generally ineffective because the oil can move under the boom at the shoreline–water interface.

Building *berms*, *ditches*, and *dams* is another way of protecting shorelines. Berms can be constructed on a beach parallel to the waterline to contain oil as it is washed ashore or below the waterline to prevent stranded oil from remobilizing. Ditches can be dug in conjunction with berms or separately to serve the same purpose. Berms or dams can be built across overwash channels to keep oil out of backshore lagoons or marshes. Berms, ditches, and dams are best used on sand shores or shorelines with impermeable fill. Ditches could also be dug on coarse shorelines, as there is usually less permeable material under the surface. As mechanical equipment is usually needed to build berms and ditches, the shoreline must be accessible and consist of firm substrates upon which to work. Berms and ditches are not recommended for use on sensitive shorelines, such as marshes and sand or mud tidal flats, due to the intrusive and potentially damaging effects of cleanup personnel and heavy equipment.

Geotextile materials and *sorbent sheets* are occasionally used to protect shorelines. This requires lead time before the oil arrives, as well as a lot of personnel and costly materials. This is usually done only where there is riprap, which is large man-made rock material for protecting shorelines, and close to man-made structures that would be difficult to clean.

Cleanup Methods

Many methods are available for removing oil from shorelines. All of them are costly and take a long time to carry out. The selection of the appropriate cleanup technique is based on the type of substrate; the depth of oil in the sediments; the amount and type of oil and its present form or condition; the ability of the shoreline to support traffic; the environmental, human, and cultural sensitivity of the shoreline; and the prevailing ocean and weather conditions. The cleanup techniques suitable for use on the various types of shoreline are listed in Table 11.1.

The primary objective of cleanup operations is to minimize the effects of the stranded oil and accelerate the natural recovery of affected areas. Obviously, a cleanup technique should be safe and effective and not be so intrusive as to cause more damage than the oil itself. In general, cleanup techniques should not be used if they endanger human life or safety, leave toxic residue or contaminate other shorelines or lower tidal areas, or kill plants and animals on the shoreline. In addition, excessive amounts of shoreline material should not be removed and the structure of the shoreline should not be changed so as to make it unstable. Saline cleaning water should not be used to clean freshwater beaches or vice versa. Finally, any technique that generates a lot of waste material should not be used. In the past, heavy equipment used on beaches resulted in thousands of tons of contaminated beach material, which then required disposal.

The length of time required to complete the cleanup is another important criteria when selecting a cleanup technique. The longer oil is on a beach, the harder it is to clean. A method that removes most of the mobile oil rapidly is much better, in many circumstances, than a more thorough one that takes weeks to carry out. Time often dictates the cleanup method used. For example, a spill on a beach where baby seals will be born in a few days would require the most rapid cleanup method rather than the method most suitable for that particular type of shoreline. Decisions on cleaning recreational beaches are often strongly influenced by seasonal usage.

Recommended Cleanup Methods

Some recommended shoreline cleanup methods are natural recovery, manual removal, flooding or washing, use of vacuums, mechanical removal, tilling and aeration, sediment reworking or surf washing, and the use of sorbents or chemical cleaning agents.

Sometimes the best response to an oil spill on a shoreline may be to leave the oil and monitor the *natural recovery* of the affected area. This would be the case if more damage would be caused by cleanup than by leaving the environment to recover on its own. This option is suitable for small spills in sensitive environments and on a beach that will recover quickly on its own

TABLE 11.1

Cleanup Techniques and Shoreline Types

Shoreline Type	Condition of the Oil	Natural Recovery	Flooding	Low-Pressure Cold Water	Low-Pressure Warm Water	Manual Removal	Vacuums	Mechanical Removal	Sorbents	Tilling/Aeration	Sediment Reworking/Surf Washing	Cleaning Agents
Bedrock	Fluid	+	+	+	◁	◁	◁		◁			◇
	Solid	+			◇	◁						+
Man-made	Fluid	+	+	+	◁	◁			◁			◇
	Solid				◇	◁						+
Boulder	Fluid	+	+	+	◁	◁			◁			◇
	Solid				◇							+
Pebble–cobble	Fluid	+	+	+		◁			◁	◁	◁	◇
	Solid					◁		◁	◁			◇
Mixed sand–gravel	Fluid	+	+	+		◁			◁	◁	◁	◇
	Solid							+	◁			
Sand beach	Fluid	+	+	+		◁		+	◁	◁	◁	
	Solid					◁		+				
Sand tidal flats	Fluid	+	+	+		◁	◁	◁	◁			
	Solid											

Mud tidal flats	Fluid	+	+	+			◁		◁	
	Solid							◁		◇
Marshes	Fluid	+	+	+					◁	
	Solid	+	+			◁				◇
Peat shorelines or low-lying tundra	Fluid	+	+	+			◁		◁	
	Solid							◁		
Mangroves	Fluid	+	+	+	◁	◁			◁	◇
	Solid			+	◇	◁				+

Note: +, acceptable method; △, suitable method for small quantities; ◇, conditional method, may only work under special circumstances.

PHOTO 11.10
Workers clean a sandy beach using the traditional methods of hand labor and bagging the oil and debris. (Photo from BP Web site: file:///C:/Graphics%20etc%20etc/photographs/Gulf%20 oil%20spill/deepwater_horizon_oil_spill. Accessed April 27, 2012.)

PHOTO 11.11
Workers hunt and collect tar balls on a sandy beach. (Photo courtesy of the U.S. National Oceanic and Atmospheric Administration [NOAA].)

such as on exposed shorelines and with nonpersistent oils such as diesel fuel on impermeable beaches. This is not an appropriate response if important ecological or human resources are threatened by long-term persistence of the oil.

Manual recovery is the most common method of shoreline cleanup. Teams of workers pick up oil, oiled sediments, or oily debris with gloved hands, rakes, forks, trowels, shovels, sorbent materials, hand bailers, or poles. It may also include scraping or wiping with sorbent materials or sifting sand to remove tar balls. Workers wear protective clothing such as splash suits,

PHOTO 11.12
Volunteers clean oiled boulders by hand. (Photo courtesy of the U.S. National Oceanic and Atmospheric Administration [NOAA].)

boots, gloves, and respirators if the oil is volatile. Material is usually collected directly into plastic bags, drums, or buckets for transfer.

Oil can be manually removed from almost any type of shoreline, although this method is most suitable for small amounts of viscous oil and surface oil and in areas inaccessible to vehicles. Although removing oil manually is a slow process, it generates less waste than other techniques and allows cleanup efforts to be focused only on oiled areas. A disadvantage is the risk of injuries to personnel from falls on slippery and treacherous shorelines.

Flooding or *washing shorelines* are common cleanup methods. Low-pressure washing with cool or lukewarm water causes little ecological damage and removes oil quickly. Warmer water removes more oil but causes more damage. High pressure and temperature cause severe ecological damage, and recovery may take years. It is preferable to leave some oil on the shoreline than to remove more oil but kill all the biota with high pressure or temperature water.

Low-pressure cool or warm water washing uses water at pressures less than about 200 kpa (50 psi) and temperatures less than about 30°C. Water is applied with hoses that do not focus the water excessively, avoiding the loss of plants and animals. Flooding is a process in which a large flow or deluge of water is released on the upper portion of the beach. Water can be applied to the beach using hoses without nozzles to reduce the impact of the spray. Sometimes a special header or pipe is used to distribute the water. Booms are then used to contain the flow and direct the recovered oil to a calm area where it can be recovered with mechanical skimmers.

Low-pressure washing and flooding are often combined to ensure that oil is carried down the beach to the water, where it can be recovered with skimmers. Washing and flooding are best done on impermeable shoreline types

PHOTO 11.13
A steel garbage bin is used to collect oily weeds and other waste from an oiled shoreline.
(Photo courtesy of the U.S. National Oceanic and Atmospheric Administration [NOAA].)

and are not useful for shorelines with fine sediments such as sand or mud. These techniques are not used on shorelines where sensitive plant species are growing.

Several sizes of *vacuum systems* are useful for removing liquid oil that has pooled or collected in depressions on beaches and shorelines. Small vacuum units are available that are specially designed for use on shorelines. The suction hose is usually applied manually to those areas where oil can be removed. Vacuum trucks used for collecting domestic waste are often used to remove large pools of recovered oil rather than for recovering oil directly on the beach. For safety reasons, vacuums are not used for oils that are volatile.

Mechanical removal involves removing the surface oil and oiled debris with tractors, front-end loaders, scrapers, or larger equipment such as road graders and excavators. There are many types of specialized beach cleaning now available. These machines can screen out tar balls and return clean sand directly to the beaches. Graders and bulldozers are sometimes used on long straight stretches of recreational sand beaches, where they quickly windrow oiled sand. Front-end loaders and backhoes are used on a variety of beaches to move oiled materials and to expose buried material as well as to remove materials recovered manually from the beach. Draglines and clamshells are used for these purposes on shorelines that are accessible from barges.

Although mechanical devices remove oil quickly from shorelines, they also remove large amounts of other material and generate more waste than other techniques, unless specialized cleaners are used. Sand and sand–gravel

PHOTO 11.14
Beach cleaning machines of this type are excellent for removing surface tar balls. (Photo from the U.S. Coast Guard Web site: http://cgvi.uscg.mil.)

PHOTO 11.15
This larger beach cleaning machine, the Sand Shark, was developed during the *Deepwater Horizon* spill and is capable of removing oil that is on or close to the surface. (Photo from Flickr. com: http://www.flickr.com/photos/bpamerica/sets. Accessed April 27, 2012.)

shorelines are best suited to this technique, as they can support mechanical equipment and are not usually damaged by the removal of material. Mechanical equipment should not be used on sensitive shorelines, shorelines with an abundance of plant and animal life, and shorelines that would become unstable if large amounts of material are removed.

Specialized shore-cleaning equipment is available that is equipped with rakes, elevating conveyors, screening devices, or spiked drums to remove oil and oily sand from sand shorelines. These devices are more selective than earth-moving equipment; they return much of the clean sand to immediately behind the device. These specialized beach cleaning devices have the potential to clean small amounts of tar balls from many miles of beaches in a single day.

Dry mixing or tilling and aeration are used to break up surface layers or to expose subsurface oil. The exposed oil can then weather naturally and degrade, and will not leach into the water. Medium to heavy oils that form crusts are also broken down and asphalt pavement buried in the beach is exposed. This work is done with farm equipment, such as ploughs, discs, rototillers, and cultivators, and construction equipment such as bulldozers or graders with rippers. The technique is suitable for sand, sand–gravel, or pebble–cobble beaches.

Sediment location, wet mixing, or *surf washing* is another method that involves the use of mechanical equipment. Oiled material is moved from the upper tidal zone, where it would remain for many years, down to the intertidal zone where the oil will be washed out by the surf. Although this is usually done with graders or front-end loaders, it can be done manually. This method is useful on sand–gravel or sand beaches where the oil has been moved above the normal high-tide line by a storm surge or very high tide. It is also used on recreational beaches that must be rapidly restored. As the oil is released from the sediment and could potentially contaminate other shorelines, this technique is not appropriate if other locations or plants and animals could be endangered.

PHOTO 11.16
This device, called the Marsh Washer, was developed during the *Deepwater Horizon* spill to clean oiled marshes. (Photo from Flickr.com:http://www.flickr.com/photos/bpamerica/sets. Accessed April 27, 2012.)

Sorbents are used in several ways in beach cleanup. In a passive role, sorbents are left in place, on or near a beach, to absorb oil that is released from the beach by natural processes and prevent it from recontaminating other beaches or contacting wildlife. Sorbent booms as well as "pom-poms" designed for heavy oil can be staked on the beach or in the water on the beach face to catch oil released naturally. This is effective but labor intensive and produces a large amount of waste material. Loose sorbents such as peat moss and wood chips are generally not used because they may sink and migrate into nonoiled areas and are difficult to recover.

Chemical cleaning agents called *beach cleaners* or **surface-washing agents** with low toxicity to aquatic organisms have been developed. These agents typically contain a surfactant and low-toxicity solvent. They act by inserting molecules between the oil and substrate, thus lessening the adhesion to the surface and releasing the oil. They are applied at low tide, allowed to soak into stranded oil, and then low-pressure washing is used to move the oil to the water where it is recovered with skimmers. High-pressure washing could cause the oil to disperse. This is considered undesirable. Approval from the appropriate environmental agencies is required before surface-washing agents can be used.

Less Recommended Cleanup Methods

Other more drastic methods, which have a greater impact on the environment, are available for cleaning shorelines. These methods may be required in certain circumstances but are less recommended than the techniques already discussed.

PHOTO 11.17
A vacuum device is used to remove excess oil from a shoreline. (Photo from Flickr.com: http://www.flickr.com/photos/bpamerica/sets. Accessed April 27, 2012.)

High-pressure cold or hot water washing have been used in the past to clean some beaches. Although very effective, especially the hot water washing, the technique removes most of the plant and animal life along with the oil, leaving the treated stretch of shoreline sterile. The technique is therefore not used on any sensitive shoreline or where the recolonization would be slow. In locations where this may be the only way to remove oil, it may be preferable to leave the oil than to remove it in this harsh manner.

Steam cleaning and *sand blasting* are similar methods in that they remove almost any type of oil from any type of shoreline, but destroy plant and animal life in the process. These methods are usually used only on man-made structures such as piles, piers, jetties, or walls.

Vegetation cutting is sometimes carried out in marshes and other areas where plant life is threatened by oiling, especially by heavy oils. Plants can be saved in this way, although traffic into marshes or other sensitive areas can cause severe damage. It is only used if it can be done without causing damage by access and if it removes oil that could recontaminate this or other areas.

In-situ burning is useful if the water level is high and the burn residue is either removed or does not suppress future plant growth. Oil will not burn on a typical beach unless the oil is pooled or concentrated in sumps or trenches with a thickness of 2 to 3 mm. In-situ burning can be sustained in marshes when the oil is pooled and when the marsh plants will burn. In fact, burning is a useful restorative method for marshes if done in spring when the water level is high so that the heat does not affect the plant roots. Burning in late summer or early fall, however, can kill much of the plant life.

PHOTO 11.18
Workers returning from hand cleaning a salt marsh. Workers are using a single path to minimize damage to the marsh. (Photo courtesy of Environment Canada.)

PHOTO 11.19
Burning an oiled salt marsh. This burn was successful and removed much of the oil. (Photo courtesy of the U.S. National Oceanic and Atmospheric Administration [NOAA].)

PHOTO 11.20
A Shoreline Countermeasures Assessment Team (SCAT) team assessing an oiled shoreline with an adjacent reedbed. (Photo courtesy of Environment Canada.)

An alternative is to flood the marsh using berms and pumps, which will raise some of the oil for burning.

Although *chemical agents* other than beach cleaners or surface-washing agents are sometimes suggested for shoreline cleanup, they should not be used as they are not effective and can cause additional problems. Dispersants increase the penetration of the oil, which makes them unsuitable for use on shorelines. Solidifiers or recovery agents do not assist with oil recovery.

Bioremediation has been used for years to remove oil from shorelines. It can be used on all types of shoreline but works only with light to medium oils that have a high saturate fraction. Generally, much of the saturate content of these oils biodegrades. The remaining oil usually consists of larger saturates, aromatics, resins, and asphaltenes, and remains to form asphalt pavement.

Applying fertilizers, in a manner similar to that described in Chapters 9 and 12, can enhance biodegradation. Fertilizers that are soluble in water will rapidly dissipate into the water, however, and can cause accelerated algal growth in nearshore waters. This can be avoided by using slowly dissolving pellets at the upper tidal reaches or oil-soluble fertilizers, although these are more expensive and less available. The effects of bioremediation are not noticeable for a long time, usually at least one season after the fertilizers have been applied. Light oils may take years to biodegrade somewhat and will never completely biodegrade. Heavy oils will only biodegrade to a small degree.

12

Oil Spills on Land

While the vast majority of oil spills in Canada occur on land (see statistics in Chapter 1), land spills are less dramatic than spills on water, and receive less attention from the media and the public. This chapter deals with the behavior of oil spilled on land and describes common methods of containment and cleanup for such spills.

Two types of land spills are discussed: (1) those that occur primarily on the surface of the land and (2) those that occur partially or totally in the subsurface. The sources and the cleanup methods differ for these types of spills. Most surface spills in Canada are the result of oil production, such as spills from pipelines and battery sites, whereas most subsurface spills are from leaking underground fuel storage tanks or pipelines. Whether on the surface or subsurface, however, each spill is unique in terms of the type of material spilled, the habitat in which the oil is spilled, its location, and the weather conditions during and after the spill.

One concern that should be borne in mind is the movement of spills to lakes and rivers. Such movement on water can spread contamination rapidly over a wide area.

Protecting human health and safety is still the top priority when cleaning oil spills on land and in the subsurface, although this is primarily an issue with some fuels, such as gasoline and diesel fuel. Minimizing long-term damage to the environment and protecting agricultural land are more often the main concerns with spills on land. This is followed by protecting nonessential uses, such as recreation.

Behavior of Oil on Land

The spreading of oil across the surface and its movement downward through soil and rock is far more complicated and unpredictable on land than the spreading of oil on water. The movement of the oil varies for different types of oil and in different habitats, and is influenced by conditions at the spill site, including the specific soil types and their arrangement, moisture conditions in the soil, the slope of the land, and the level and flow rate of the groundwater. Other factors, which vary in different habitats, are the presence of vegetation and its type and growth phase, the temperature, the presence

TABLE 12.1

Properties of Different Oils and Their Effect on the Soil Environment

Petroleum	Plant Toxicity	Water Threat	Viscosity	Adhesion	Penetration	Degradation
Gasoline	5	5	1	1	5	4
Diesel fuel	2	3	2	2	4	1
Light crude	4	4	3	3	3	2
Heavy crude	3	2	4	4	2	3
Bunker fuel	1	1	5	5	1	5

Note: Lower numbers indicate more favorable conditions to the environment and faster recovery after a spill.

PHOTO 12.1
The ground after the rapid passage of several tons of hot Bunker C oil. This ground was cleaned by the removal of the surface and replacement of soil and grass. (Photo courtesy of Environment Canada.)

of snow and ice, and the presence of microfeatures, such as rock outcrops. Some properties of different oils and their effects on the soil environment are shown in Table 12.1.

The basic types of soil to consider in relation to oil spills are sand/gravel, loam, clay, and silt. "Soil" is defined as the loose unconsolidated material located near the surface, whereas "rock" is the hard consolidated material, that is, bedrock, usually found beneath the soil. Most soils consist of small fragments or grains that form openings or pores when compacted together. If these pores are sufficiently large and interconnected, the soil is said to be "permeable," and oil or water can pass through it. Sand is the most permeable type of soil. Materials such as clay, silt, or shale are termed "impervious" as they have extremely small, poorly interconnected pores and allow only limited passage of fluids. Soils also vary in terms of long-term retentivity. Loam tends to retain the most water or oil due to its high organic content.

As most soils are an inhomogeneous mixture of these different types of soil, the degree of spreading and penetration of oil can vary considerably in a given location. The types of soil are often arranged in layers, with loam on top and less permeable materials such as clay or bedrock underneath. If rock is fractured and contains fissures, oil can readily pass through it.

The oil's ability to permeate soils and its adhesion properties also vary significantly. Viscous oils, such as bunker fuel oil, often form a tarry mass when spilled and move slowly, particularly when the ambient temperature is low. Nonviscous products, such as gasoline, move in a manner similar to water in both summer and winter. For such light products, most spreading occurs immediately after a spill.

Crude oils have intermediate adhesion properties. In an area with typical agricultural loam, spilled crude oil usually saturates the upper 10 to 20 cm of soil and rarely penetrates more than 60 cm. Generally, the oil only penetrates to this depth if it has formed pools in dry depressions. If the depressions contain water, the oil may not penetrate at all.

Movement of Oil on Land Surface

Both the properties of the oil and the nature of the soil materials affect how rapidly the oil penetrates the soil and how much the oil adheres to the soil. For example, a low viscosity oil penetrates rapidly into a dry porous soil such as coarse sand and therefore its rate of spreading over the surface is reduced.

When oil is spilled on land, it runs off the surface in the same direction and manner as water. The oil continues to move horizontally down gradient until either blocked by an impermeable barrier or all the oil is absorbed by the soil. The oil will also sink into any depressions and penetrate into permeable soils.

The process whereby oil penetrates through permeable soils is shown in Figure 12.1. The bulk of the oil moves downward through permeable material under the influence of gravity until either the groundwater or an impermeable layer stops it. It then moves down gradient along the top of the impermeable layer or groundwater until it encounters another impermeable barrier or all the product is absorbed in the soil. Once in contact with the water-soluble material, the oil dissolves into it and is transported away with the groundwater. Oils and fluids can flow along the top of the groundwater and reappear much later in springs or rivers.

The descending oil is often referred to as a "slug" of oil. As the slug moves through the soil, it leaves material behind that adheres to the soil. This depends on the adhesion properties of the spilled product and the nature

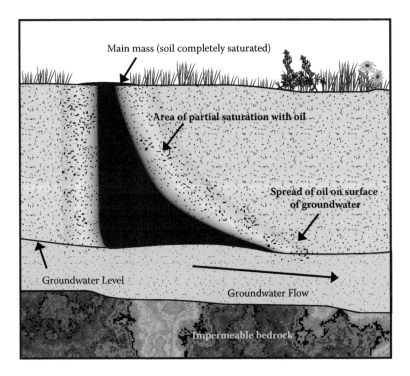

FIGURE 12.1
Penetration of oil into soil.

of the soil. More of the adhered oil is moved downward by rainfall percolating through the soil. The rainwater carries dissolved components with it to the water table. The movement of the oil will be greatest where the water drainage is good.

Movement of Oil in the Subsurface

Regardless of its source, oil released into the subsurface soil moves along the path of least resistance and downward, under the influence of gravity, as shown in Figure 12.2. Oil often migrates toward excavated areas, such as pipeline trenches, filled-in areas around building foundations, utility corridors, and roadbeds. Such areas are often filled with material that is more permeable or less compacted than the material removed during the excavation.

The oil may continue to move downward until it reaches the groundwater or another impermeable layer. If the soil is absorptive and capillary action occurs, however, the oil can also move upward and even reappear at

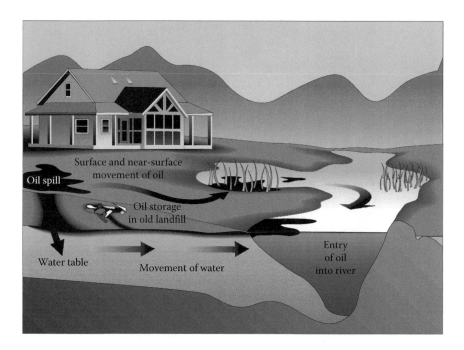

FIGURE 12.2
Subsurface movement of oil.

the surface, sometimes as far as a kilometer away from the spill. This is what happens when pipeline spills appear at the surface of the trench in which the pipeline is laid.

Habitats and Ecosystems

As the effects of oil and its behavior vary in different habitats, cleanup techniques and priorities are tailored to the habitat in which the spill takes place. Returning the habitat as much and as quickly as possible to its original condition is always a high priority when cleaning up oil spills.

It is important to note that each site may be very inhomogeneous in terms of its vegetation, soil types and soil profile, and how the oil behaves in or affects each component of the soil. Furthermore, the amount of time it takes for the vegetation to grow back naturally differs widely from one habitat to another. The estimated amount of time for surface vegetation to recover in various oiled habitats is shown in Table 12.2. Residual amounts of oil remain in some habitats for many years or even decades.

When spills occur in the *urban habitat*, protecting human health and safety, and quickly restoring the land use are top priorities. Environmental

PHOTO 12.2
An aerial photograph of a recent pipeline spill. The spilled oil is running to lower land and to
the location of some water collection areas.

TABLE 12.2

Estimated Recovery Times in Various Habitats

Habitat	Recovery Time without Cleanup (Years)	Recovery Time with Minimal Cleanup (Years)	Recovery Time with Optimal Cleanup (Years)
Urban	1 to 5	1	<1
Roadside	1 to 5	1	<1
Agricultural land	2 to 10	1 to 5	1 to 2
Dry grassland	1 to 5	1 to 3	1 to 2
Forest	2 to 20	2 to 5	1 to 3
Wetland	5 to 30	3 to 20	2 to 10
Taiga	3 to 20	2 to 10	2 to 8
Tundra	3 to 15	2 to 10	2 to 8

considerations are generally not important as endangered species or ecosystems are not often found in the urban habitat. The urban environment usually includes a range of ecosystems, from natural forest to paved parking lots. Thus, a spill in an urban area often affects several ecosystems, each of which is treated individually.

The *roadside habitat* is similar to the urban one in that restoring the use and surficial appearance of the land is given top priority. Roadside habitats are varied and include all the other ecosystems. The roadside habitat is different from the urban one, however, in that it is exposed to many emissions and is not generally viewed as a threatened or sensitive environment.

On *agricultural land*, the priority in cleaning up oil spills is to restore land use, for example, crop production. In this habitat, oil is more likely to penetrate deeply into the subsurface as plowing the fields creates macropores that petroleum products and crude oils can rapidly penetrate. As oil penetrates deeper into dry agricultural land, the danger of groundwater contamination is greater than in other habitats.

On mineral soils, however, oil can make the soil nonwettable, so that water runs off rather than soaking into the soil. This causes a water shortage, which can result in poor rehabilitation in the area. The opposite occurs in low-lying sites or poorly drained soils where water fills the macropores of the soil, but is not absorbed into the soil itself because of the presence of the oil. This excludes air from the soil and the site becomes difficult to treat or cultivate and anaerobic conditions quickly develop.

Anaerobic conditions and restricted plant growth can also develop when oil on the surface weathers and forms an impermeable crust, which again reduces the air exchange. Recovery is affected by the amount of oil spilled on a given area. Lightly oiled soil recovers much faster than a heavier oiled area, as the soil is not completely saturated and both air and water can still

PHOTO 12.3
This photo illustrates the complexity of a large pipeline spill into a low area.

penetrate. Residual oil in the soil can also slow recovery by inhibiting seed germination.

Dry grassland is similar to agricultural land in that the priority for cleanup is restoring the soil so that the crop, in this case grass, can continue to grow. The surface of the grassland is often less permeable than agricultural land. Once the surface is penetrated, however, the substrate may be permeable and groundwater can be affected. Dry grassland recovers quickly from spills if the oil runs off or if the excess oil is removed without too much surface damage. The presence of dead vegetation is viewed as a symptom, not a problem. When excess oil is removed, replanting and fertilization can speed recovery of an oiled grassland. As with agricultural land, oil on the surface of grassland can sometimes weather and form an impermeable crust that reduces air exchange and causes anaerobic conditions.

Unlike most habitats, the *forest* has two distinct levels of vegetation: (1) low-lying vegetation such as shrubs and grasses, and (2) trees. The low-lying vegetation is much more sensitive to oiling than trees, but is much easier to replant and recovers much faster. Most species of trees are not seriously affected by light oil spills. If enough oil is spilled to affect the tree's roots, most trees will be killed and the forest will not recover fully for decades. It is therefore very important to rapidly remove excess oil that has not yet been absorbed by the soil.

If a forest has mineral soil, the oil can make it nonwettable so that water runs off the soil rather than soaking in. In low-lying sites or forests with poorly drained soil, the opposite occurs. Water fills the macropores of the soil but not the soil itself because of the presence of the oil. This excludes air from the soil and the site does not revegetate quickly. Oil on the surface of

PHOTO 12.4
This area was impacted by an oil spill 25 years before this photo was taken. The area was treated and rototilled. Vegetation has never been reestablished. (Photo courtesy of Environment Canada.)

forest soils can weather and form an impermeable crust that also reduces air exchange or restricts the growth of plants. Due to the presence of large trees, the forest is far more difficult to access and treat than most other habitats.

Wetlands are the habitat most affected by oil spills because they are at the bottom of the gravity drainage scheme. Usually, oil cannot flow out of a wetland system and oil from other areas flows into the system. Although there is a variety of wetlands, oil tends to collect in all of them, creating anaerobic conditions that slow oil degradation. Wetlands are also extremely sensitive to physical disturbance as many plants in this habitat propagate through root systems. If these root systems are damaged by the oil or the cleanup process, it takes years or even decades for the plants to grow back. Wetlands are the habitat of many species of birds and fish as well as other aquatic resources. Wetlands are difficult to access and to clean.

Taiga, which is characterized by coniferous trees and swampy land, generally forms the transition between northern forests and the tundra farther to the north. It is either underlain by permafrost or has a high water table. Many of the plants propagate through root systems and are highly sensitive to physical disturbance. Over a period of time, heavy loadings of oil will kill the coniferous trees. Oil on the surface of the taiga can weather and form an impermeable crust that reduces the air exchange and restricts plant growth. Degradation of remaining oil is slow in this habitat, which takes a long time to recover. The presence of trees and the high moisture level make the taiga more difficult to access and clean than most other habitats.

Tundra is a far northern habitat, characterized by low plant growth and no trees. Tundra is underlain by permafrost, which is generally impermeable to oil. Vegetation on the tundra grows in tufts that are generally grouped into polygons. Oil spilled on the surface drains into the spaces between the tufts

PHOTO 12.5
A marsh drainage channel impacted by a Bunker C oil spill. Workers are manually cleaning both the drainage channel and the vegetation. (Photo courtesy of Environment Canada.)

and polygons, and eventually kills the vegetation. Without the layer of vegetation, the permafrost melts and serious land damage results. Degradation of remaining oil on the tundra is very slow and could take hundreds of years.

In all the more sensitive habitats, which include the forest, the taiga, and the tundra, the priority for cleanup operations is to remove the excess oil as rapidly as possible and without causing physical damage.

Cleanup of Surface Spills

When dealing with oil spills on land, cleanup operations should begin as soon as possible. It is important to prevent the oil from spreading by containing it and to prevent further contamination by removing the source of the spill. It is also important to prevent the oil from penetrating the surface and possibly contaminating the groundwater.

PHOTO 12.6
A land spill being cleaned up using a vacuum truck. (Photo courtesy of Environment Canada.)

Berms or *dikes* can be built to contain oil spills and prevent oil from spreading horizontally. Caution must be exerted, however, that the oil does not back up behind the berm and permeate the soil. Berms can be built with soil from the area, sandbags, or construction materials. Berms are removed after cleanup to restore the area's natural drainage patterns. *Sorbents* can also be used to recover some of the oil and to prevent further spreading. The contaminated area can sometimes be *flooded* with water to slow penetration and possibly float oil to the surface, although care must be taken not to increase spreading and to ensure that water-soluble components of the oil are not carried down into the soil with the water. *Shallow trenches* can be dug as a method of containment, which is particularly effective if the water table is high and oil will not permeate the soil. Oil can either be recovered directly from the trenches or burned in the trenches. After the cleanup, trenches are filled in to restore natural water levels and drainage patterns.

There are a variety of methods for cleaning surface oil spills on land, with the method used depending on the habitat in which the spill occurs. The various cleanup methods that should be used in the different habitats are shown in Table 12.3.

Natural Recovery

Natural recovery is the process of leaving the spill site to recover on its own. This method is sometimes chosen for extremely sensitive habitats such as wetlands, taiga, and tundra, and is always done after the excess oil has been removed from the site. In these cases, the excess oil that can be recovered is removed using techniques that do not disturb the surface or physically damage the environment. This is important as it can take years for wetlands or tundra to recover from vehicular traffic. In habitats such as wetlands and taiga where the vegetation propagates through root systems, more damage can be done by the cleanup operation than by the oil.

Removal of Excess Oil

Any excess oil that can be recovered without causing physical damage to the environment is always removed from a spill site using techniques that do not disturb the surface. If excess oil on the surface is not removed quickly, the oil can penetrate into the soil, contaminating the groundwater and destroying vegetation.

Suction hoses, pumps, vacuum trucks, and certain *skimmers and sorbents,* both natural and synthetic, are generally effective in removing excess oil from the surface, especially from ditches or low areas. The use of sorbents can complicate cleanup operations, however, as contaminated sorbents must be appropriately disposed. Sorbents are best used to remove the final traces of oil from a water surface. Any removal of surface soil or vegetation also entails replanting and fertilization.

TABLE 12.3

Cleanup Methods for Surface Land Spills

Habitat	Removal of Excess Oil	Natural Recovery	Manual Oil Removal	Mechanical Oil/ Surface Removal	Enhanced Biodegradation	In-Situ Burning	Hydraulic Measures
Urban	√	+	√	+	+	◊	√
Roadside	√	+	√	+	+	+	+
Agricultural land	√	+	√	+	+	+	√
Grassland	√	+	√	+	+	+	+
Forest	√	√	√	+	+	◊	√
Wetland	√	√	√	×	+	√	+
Taiga	√	√	√	×	+	+	+
Tundra	√	√	√	×	+	+	+

Note: √, acceptable or recommended; +, can be used under certain circumstances; ×, should not be used; ◊, only marginally applicable.

PHOTO 12.7
An interceptor trench is used to catch oil flowing on the groundwater.

Manual removal of oil involves removing oil and often highly oiled soil and vegetation with shovels and other agricultural tools. This is always followed by fertilization, selective reseeding, or transplanting plugs of vegetation from nearby unaffected areas. This form of removal is labor intensive and can severely damage the surface, especially in sensitive environments.

Mechanical recovery equipment, such as bulldozers, scrapers, and front-end loaders, can cause severe and long-lasting damage to sensitive environments. It can be used in a limited capacity to clean oil from urban areas, roadsides, and possibly on agricultural land. The unselective removal of a large amount of soil leads to the problem of disposing of the contaminated material. Contaminated soil must be treated, washed, or contained before it can safely be disposed in a landfill site. This can cost thousands of dollars per ton.

Other Cleanup Methods

Enhanced biodegradation is another possible method for cleaning spills on land. Certain portions of oil are biodegradable and the rate of biodegradation can sometimes be accelerated as much as tenfold by the proper application of fertilizers.

The amount of degradation varies with the type of oil. Diesel fuel will largely degrade on the land surface, whereas Bunker C will only slightly degrade. Under ideal conditions and using fertilizers to enhance degradation, however, it can still take from 3 to 100 years for more than half of an oil to degrade, even if this oil is biodegradable. During this time, some of the oil will be removed by other processes such as evaporation or simply by movement.

Scientists are now exploring the use of plants and their associated microorganisms for remediation, which is called *phytoremediation*. This is a low-cost process that is proving effective for a wide variety of contaminants, including petroleum hydrocarbons. It can be used in combination with other remediation technologies and may prove useful in the future for treating oiled soils or wetlands. It takes several years to remediate a site and cleanup is limited to the depth of the soil within reach of the plant's roots.

In-situ burning has been used for several years to deal with oil spills on land. This technique removes oil quickly and without disturbing the area extensively, although it does damage or kill shrubs and trees. The heat from burning can also destroy propagating root systems and change the soil's properties. In addition, it can leave a hard crust of residual material that inhibits plant growth and changes natural water levels and drainage patterns.

These disadvantages can be overcome in some habitats. Some areas can be flooded before burning to minimize the effect of the heat and to remove oil by floating it out of the ground. Crust formation can be avoided by physically removing residue after a burn. On wetlands and in areas with high water levels, sorbents can be used to remove residues left after burning to ensure that they do not coat plants or soil after water levels fall. In marshes, burning is best done in spring when the water table is high.

Hydraulic measures, such as flooding and cold or warm water sprays, can be used to deal with land spills, although they are only effective in limited circumstances. *Flooding* an area where the oil is not strongly retained can cause the oil to rise to the water's surface where it can subsequently be removed using skimmers or suction devices, or by burning. This is effective in areas where the water table is high or the top layer of ground is underlain by impermeable material. Flooding may not work on soils that are high in organic material, however, as they strongly retain oil. *Cold or warm water sprays* can be used to clean oil from hard surfaces. Catchment basins and interceptor trenches are built to capture the released oil, which is then skimmed or pumped from the trenches.

A number of other techniques have been tried for cleaning oil spills on land, with varying degrees of success. *Tilling or aeration of soil* is done to break up the crust surface and reaerate the soil. In areas where vegetation propagates by root systems, however, tilling kills all plants and destroys the potential for regrowth. Tilling oil into the soil can actually slow natural degradation because the soil can become anaerobic when it recompresses. *Vegetation cutting* is useful only if there is a risk that oil on vegetation could recontaminate other areas. Many plants cannot survive cutting, however, and growth is not reestablished in the area. To date, there are no effective *chemical agents* for cleaning oil spills. Surfactant agents can actually increase oil penetration into the soil and could result in the more serious problem of groundwater contamination.

All cleanup methods include *site restoration*, which involves returning the site as closely as possible to the prespill conditions. The drainage pattern of the site is restored by removing dikes, dams, and berms, and filling in ditches or drains. It may be necessary to replace any soil that was removed and to revegetate the site by fertilizing, reseeding, or transplanting vegetation from nearby.

Cleanup of Subsurface Spills

Oil spills in the subsurface are much more complicated and expensive to clean than those on the surface, and the risk of groundwater contamination is greater. Spills in the subsurface can be difficult to locate and without knowledge of the geology of the area, it can be difficult to predict the horizontal and downward movement of the oil. The first step is typically to engage a hydrogeologist to map and assess both the subsurface and the oil location with respect to the soil geology.

PHOTO 12.8
An aboveground pipeline leaks into a taiga environment. This may take many years to clean up. (Photo courtesy of Environment Canada.)

In terms of countermeasures, the oil must be contained and its horizontal and downward movement stopped or slowed. Containment methods are difficult to implement and may cause physical damage to the site. Digging an *interceptor trench* can be effective in reducing horizontal spread. Such trenches are filled in after the cleanup operation to restore the natural drainage patterns of the land. Another method is to place "walls" around the spill source to stop its spreading. These can be "slurry walls" consisting of clay or cement mixtures that solidify and retain the oil, or solid sheets of steel or concrete can be positioned to retain the oil.

Once the subsurface spill is contained, there are a number of cleanup methods that can be used. The most appropriate method for a particular spill depends on the type of oil spilled and the type of soil at the site, as shown in Table 12.4.

Hydraulic measures for cleaning subsurface spills include flooding, flushing, sumps, and French drains. These methods are most effective in permeable soil and with nonadhesive oils. They all leave residual material in the soil that may be acceptable, depending on the land use. *Flooding* is the application of water either directly to the surface or to an interceptor trench in order to float out the oil. Flooding is effective only if the spilled oil has not already been absorbed into the soil, if sufficient water can be applied to perform the function, and if the oil is not accidentally moved into another area. *Flushing* involves the use of water to flush oil into a sump, recovery well, or interceptor trench. Placement of a *sump* or a deep hole is only effective for a light fuel in permeable soil above an impermeable layer of soil. A *French drain*

TABLE 12.4

Cleanup Methods for Subsurface Spills

Product Type in Soil Type	Hydraulic Measures	Interceptor Trench	Soil Venting	Soil Excavation	Recovery Wells
Gasoline in sand or mixed till	√	+	√	+	+
Gasoline in loam or clay	+	+	◊	+	√
Diesel fuel in sand or mixed till	√	+	◊	+	+
Diesel fuel in loam or clay	√	+	◊	+	√
Light crude in sand or mixed till	√	+	+	+	+
Light crude in loam or clay	+	+	◊	+	+
Heavier oils in sand or mixed till	+	+	◊	+	◊
Heavier oils in loam or clay	+	+	◊	+	◊

Note: √, acceptable or recommended; +, can be used under certain circumstances; ◊, only marginally applicable.

is a horizontal drain placed under the contamination, from which the fuel and often water are pumped out. Although effective in permeable soils, they are expensive and difficult to install.

Interceptor trenches are ditches or trenches dug down gradient from the spill, or in the direction in which the spill is flowing, to catch the flow of oil. They are placed just below the depth of the groundwater so that oil flowing on top of the groundwater will flow into the trench. Both water and oil are removed from the trench to ensure that flow will continue. Interceptor trenches are effective if the groundwater is very close to the surface and the soil above the groundwater is permeable.

Soil venting is done to remove vapors from permeable soil above a subsurface spill. This is effective for gasoline in warm climates and for portions of very light crude oils. Other oils do not have a high enough rate of evaporation to achieve a high recovery rate. Venting can be passive, in which vapors are released as a result of their own natural vapor pressure, or active, in which air is blown through the soil and/or drawn out with a vacuum pump. The fuel vapors are subsequently removed from the air to prevent air pollution. Soil venting is also done to enhance biodegradation.

Excavation is a commonly used technique for cleaning subsurface spills, especially in urban areas where human safety is an issue. Vapors from gasoline can travel through the soil and explode if ignited. These vapors can also penetrate houses and buildings, forcing evacuation of the area. To prevent these situations, contaminated soils are often quickly excavated and treated or packaged for disposal in a landfill. Excavation may not always be possible, however, depending on the size of the spill and prevailing conditions at the site.

Recovery wells are frequently used in cleaning subsurface spills. The well is drilled or dug to the depth of the water table so that oil flowing along the top of the water table will also enter the well. The water table is sometimes lowered, by pumping, to speed the recovery of the oil and to increase the area of the collection zone. The oil is recovered from the surface of the water by a pump or a specially designed skimmer.

Other methods are constantly being proposed or tried for cleaning subsurface spills. One such method is *biodegradation in-situ*, although its effectiveness is very much restricted by the availability of oxygen in the soil and the degradability of the oil itself. An adaptation of the venting method has been used to try to solve the oxygen problem. So far, however, biodegradation methods have not been rapid enough to be an acceptable solution. *Chemical agents* have also been proposed for cleaning subsurface spills, although most of them actually make the problem worse. For example, surfactants can release the oil from soil but then render that same oil dispersible in the groundwater.

If the groundwater does become contaminated, it is pumped to the surface and treated to remove the dissolved components. Common treatment methods include reverse osmosis and carbon filtration. Groundwater treatment is expensive and generally involves a lengthy process before contamination levels are below acceptable standards.

13

Effects of Oil Spills on the Environment

Oil spills have many adverse effects on the environment. Oiled birds are one frequent and highly publicized outcome of oil spills, but there are many other less obvious effects such as the loss of phytoplankton and other microscopic forms of life. These effects are varied and influenced by a number of factors. This chapter reviews the effects of oil on the environment and touches on how damage from oil spills is assessed.

Biological Effects

Before discussing the actual effects of an oil spill on various elements of the environment such as birds and fish, the types of effects will be discussed. Toxic effects are classified as chronic or acute, which refers to the rate of effect of toxin on an organism. *Acute* means toxic effects occur within a short period of exposure in relation to the life span of the organism. For example, acute toxicity to fish could be an effect observed within 4 days of a test. The toxic effect is induced and observable within a short time compared to the life span of the fish. *Chronic* means occurring during a relatively long period, usually 10% or more of the life span of the organism. It might take a significant portion of the life span for a chronic toxic effect to be observable, although it could have been induced by exposure to a substance that was normally acutely toxic. Chronic toxicity refers to long-term effects that are usually related to changes in such things as metabolism, growth, reproduction, or ability to survive.

The effects of exposure to a toxic substance can be lethal or sublethal. Lethal exposure is often described in terms of the concentration of the toxicant that causes death to 50% of a test population of the species within a specified period of exposure time. This is referred to as the LC_{50}. For example, tests of the effects of various crude oils on *Daphnia magna*, the water flea, show that 5 to 40 mg/L of the oil for a period of 24 hours is lethally toxic. The units of milligrams per liter (mg/L) are approximately equivalent to parts per million (ppm). Sublethal means detrimental to the test organism but below the level that directly causes death within the test period. For example, it has been found that a concentration of 2 ppm of crude oil in water causes disorientation in *Daphnia magna* when the organism is exposed for 48 hours.

PHOTO 13.1
A photo of a large number of dead clams on a beach. This was the result of a large diesel fuel spill. (Photo courtesy of the U.S. National Oceanic and Atmospheric Administration [NOAA].)

Oil can affect animals in many ways, including changing their reproductive and feeding behavior, and causing tainting and loss of habitat. Oiling of more highly developed animals such as birds may result in behavioral changes, such as failure to take care of their nests, resulting in the loss of eggs. Even a light oiling can cause some species of birds to stop laying eggs altogether.

Feeding behavior might also change. Seals sometimes react to oiling by not eating, which compounds the negative effects of the oil. The loss of an organism's habitat due to oiling can be as harmful as direct oiling because alternative habitats may not be available and the animal can perish from exposure or starvation.

Finally, tainting becomes an issue with fish and shellfish after an oil spill. Tainting occurs when the organism takes in enough hydrocarbons to cause an unpleasant, oily taste in the flesh. These organisms are unsuitable for human consumption until this taste disappears, which could take up to a year after the spill. After an oil spill, food species in the area are often tested using both chemical methods and a taste panel, and the area is sometimes closed to commercial fishing as a precaution.

Oil can enter organisms by several exposure routes: physical exposure, ingestion, absorption, and through the food chain. Animals or birds can come into direct contact with oil on the surface of water, on shorelines, or on land. The effects from this form of exposure are usually quite different than the effects of direct ingestion. Ingestion occurs when an organism directly consumes oil, usually by accident as in the case of birds when oil is ingested as they preen or groom their feathers.

Absorption of volatile components of oil is a common method of exposure, especially for plants and sessile (immobile) organisms, although it also occurs in birds and mammals. Fresh crude oil has a relative abundance of volatile compounds such as benzene and toluene, which are readily absorbed through the skin or plant membrane and are toxic to the organism.

After a spill, organisms can also be exposed to oil that passes through several organisms via the food chain. Bioaccumulation, the accumulation of toxins in the flesh, rarely occurs since the components of oil are generally metabolized by the receiving organism.

The effects of oil on the flora and fauna of a region are influenced by many factors, including the sensitivity of an organism, its recovery potential, its tendency to avoid an oil spill, its potential for rehabilitation, and the particular life stage of the organism.

Sensitivity describes how prone an organism is to the oil and any effects resulting from the oil exposure. It varies with such factors as species, season, and weather conditions. Often sensitivity maps used by spill cleanup crews include information on the vulnerability of local species to oil spills.

Recovery potential refers to the ability of organisms or ecosystems to return to their original state, or the state they were in before the spill event. Recovery time varies from days to years. For example, the ecosystem of a rocky shoreline can recover from an oil spill within months as organisms from unoiled areas can move in and restore the population.

Avoidance is another response to oil spills. Some species of fish, seals, and dolphins will avoid surface slicks and move to unoiled areas. Some birds, however, are attracted to oil slicks, mistaking them for calm water.

Another factor that influences the effects of oiling is the potential for rehabilitation of oiled animals. Birds, otters, and seals are often cleaned, treated, and returned to the environment. Many species cannot be rehabilitated, however, as they are difficult to catch and the stress of being caught and kept in captivity may be worse than the effects of oiling.

And finally, the effects of oil on any species often depend on the age or life stage of the organism. For example, juveniles of a species are often much more sensitive to oiling than the adults. Seals, for example, are much more sensitive to oiling when they are molting.

Aquatic Environments

The sea includes a wide variety of ecosystems, species, and habitats. When looking at the effects of oil spills, it is convenient to divide these into fish, plankton, benthic invertebrates, epontic organisms, marine mammals, intertidal and shoreline organisms, marine plants, and special ecosystems.

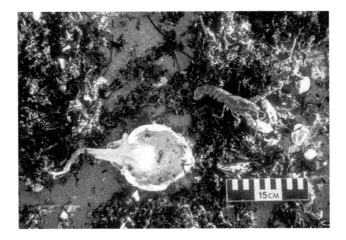

PHOTO 13.2
A number of dead lobsters and a dead skate from a diesel spill lie on the shoreline. It is esti-
mated that about 12 million lobsters died in this spill. (Photo courtesy of the U.S. National
Oceanic and Atmospheric Administration [NOAA].)

Many freshwater biota respond to oil in a manner similar to their saltwater counterparts. Although freshwater studies have not been as extensive as
those for marine situations, few differences were noted. Although oil is less
soluble in freshwater, this is largely offset by the fact that many freshwater
bodies are much shallower than oceans. A spill in a slough or pond can easily result in toxic concentrations throughout the entire water column. The
high water circulation in most rivers, however, means that hydrocarbon concentrations in the water are quickly diluted.

Fish

There is often concern about the effect of oil on fish, from both an environmental and a commercial viewpoint, because fish are an important food
source. Both pelagic (midwater) and demersal (bottom-dwelling) fish are
exposed to toxicity primarily through aromatic hydrocarbons in the water
column. The concentration of aromatic hydrocarbons in oils varies as does
the toxicity of the different aromatic compounds. Although lethal concentrations are rarely found in open seas, such concentrations can occur in
confined waters, such as bays and estuaries, directly under or near spills.
Whereas high concentrations of oil have caused massive fish mortality in
some incidents, fish are more typically exposed to sublethal concentrations
of hydrocarbons. Some concentrations of hydrocarbons that are lethal to various aquatic species, both freshwater and saltwater, are listed in Table 13.1.

The age of a fish is very important in terms of its sensitivity to hydrocarbons, with adult fish tending to be less sensitive than juveniles. For example,
tests have shown that adult salmon are 100 times less sensitive to aromatic

TABLE 13.1

Aquatic Toxicity of Water-Soluble Fractions of Common Oils

Oil Type	Specific Type	Species	Common Name	LC$_{50}$ (mg/L)[a]	Time (hr)
Gasoline		*Daphnia magna*	Water flea	20 to 50	48
		Artemia	Brine shrimp	5 to 15	48
			Rainbow trout larvae	5 to 7	48
Diesel fuel		*Daphnia magna*	Water flea	1 to 7	48
		Artemia	Brine shrimp	1 to 2	48
			Rainbow trout larvae	2 to 3	48
Light crude	Alberta sweet mixed blend	*Daphnia magna*	Water flea	6 to 12	48
		Artemia	Brine shrimp	10 to 20	48
			Rainbow trout	10 to 30	96
			Frog larvae	3	96
	Arabian light	*Daphnia magna*	Water flea	10	48
Medium crude	Cook inlet	*Fundulus*	Fish	50	96
			Scallops	2	96
			Salmon	2	96
			Crab	1	96
Heavy Crude	Arabian heavy	*Daphnia magna*	Water flea	5 to 8	48
Intermediate Fuel oil	IFO-180	*Daphnia magna*	Water flea	1 to 8	48
		Artemia	Brine shrimp	0.8 to 4	48
			Rainbow trout larvae	2	96
Bunker C		*Daphnia magna*	Water flea	0.5 to 5	48
		Artemia	Brine shrimp	0.3 to 3	48
			Rainbow trout larvae	2	96

[a] LC$_{50}$ is the lethal toxicity to 50% of the test population at the water concentration, specified in milligrams per liter (mg/L), which is approximately the equivalent of parts per million.

hydrocarbons than juvenile salmon. In turn, the juveniles are 70 times less sensitive than the salmon eggs. Several studies have shown that fish larvae or newly hatched fish are often more sensitive than fish eggs.

Other variables that determine the toxicity of hydrocarbons are the salinity and temperature of the water, the abundance of food, and the general health of the species.

Oil exposure can cause a range of physiological and pathological changes in fish, some of which are temporary and are not a risk to health or survival.

Other sublethal effects such as the disruption of growth or decreased assimilation of food may affect long-term survival. Some of the effects noted on fish such as eye cataracts, structural changes of fins, and loss of body weight may be related to the stress of exposure and not directly to the hydrocarbons.

In controlled tests, some adult fish species avoided oil slicks on the surface or dissolved hydrocarbons in the water, but this behavior has not been observed in open water spills. The conclusion is that at least some species would avoid an oil spill on open water if they can escape it.

There is concern that oil spills could disrupt the spawning behavior of anadromous species, such as salmon, that live their adult lives in saltwater but return to freshwater streams to spawn. Tests have shown that although salmon will sometimes avoid oil on open water, the exposure to oil may not badly disrupt their "homing instinct," as they tend to continue on to their freshwater home streams. Experience in actual spills has not been recorded.

PHOTO 13.3
A spill cleanup worker deals with an oiled brown pelican. (Photo from the U.S. Coast Guard Web site: http://cgvi.uscg.mil.)

There is no evidence that hydrocarbons bioaccumulate in fish or any other aquatic species. Rather, fish and other aquatic organisms tend to "depurate" or lose hydrocarbons that they have taken up. This process can take as long as 1 year from the time fish are exposed to high, sublethal concentrations of hydrocarbons until the level is below detection.

Fish species that live or spend time close to the water surface, the shore, or the seafloor are the most vulnerable to oil spills. Species with eggs or larvae that stay close to the surface and those that feed on organisms near shorelines or on the sea bottom are at greatest risk. Fish that spend most of their life stages in open waters are rarely at risk.

Plankton

Plankton are small plants and animals that live in the water and include phytoplankton and zooplankton. *Phytoplankton* are microscopic plants such as algae and diatoms that live in the top layer of the water as they depend on light for photosynthesis. *Zooplankton* are microscopic animals that feed primarily on phytoplankton. Plankton are important because they are at the bottom of the aquatic food chain. Thus, oil ingested or absorbed by plankton is passed higher up the food chain until it is finally ingested by fish and mammals.

Both phytoplankton and zooplankton vary in their sensitivity to whole oil or hydrocarbons in the water column. Plankton are killed by relatively low concentrations of oil but are present in such numbers that lost individuals are replaced quickly with little detectable disturbance. Plankton also tend to depurate low concentrations of hydrocarbons within days. Some sublethal effects of oil on zooplankton include narcosis, reduced feeding, and disruption of normal responses to light.

PHOTO 13.4
A seal colony that has been affected by an oil spill. More than 7000 seals in this colony perished as a result of this oiling.

Benthic Invertebrates

The benthos refers to the environment on the bottom of bodies of water and includes plankton, fish, and other species already discussed. Benthic invertebrates that dwell on or in the seafloor include bivalves such as clams, polychaete worms, and many mobile crustaceans such as crabs, shrimp, lobster, and amphipods.

Benthic invertebrates are generally divided into two groups: benthic infauna that resides within the bottom sediments and benthic epifauna that live mostly on the top of the sediments. Mobile forms include the slow-moving starfish, gastropods, and sea urchins. Fast-moving species include amphipods and isopods, tiny invertebrates that are an important food source for fish, bottom-feeding whales, and some species of birds, which thereby pass contamination up through the food chain. These species have the advantage of being able to avoid contaminated areas or to quickly recolonize them, whereas it can take years for sessile (or immobile) organisms to recolonize an area.

Benthic species can be killed when large amounts of oil accumulate on the bottom sediments. This can occur as a result of sedimentation, which is the slow downward movement of oil with or without sediment particles attached, or by precipitation down with or in plankton. Sometimes the oil itself is heavy enough to sink. High concentrations of hydrocarbons in the water column have killed epifauna, particularly in shallow areas or nearshore environments.

Several trends have been noted in the response of benthic invertebrates to oil. Larval stages are much more sensitive than adults, organisms undergoing molting are very sensitive, and less mobile species are more affected. Sublethal hydrocarbon concentrations cause narcosis (deathlike appearance when the organism is not actually dead) in most benthic invertebrates.

PHOTO 13.5
A dead fish as a result of oiling in a lake. (Photo courtesy of Environment Canada.)

The invertebrates often recover, although they may be more vulnerable to predators or to being swept away by currents. In 1996, a spill of diesel fuel off the east coast of the United States dispersed naturally into a nearshore region. The high level of hydrocarbons caused by dispersion narcotized or killed millions of lobsters, which were carried onto the shore where those still alive were killed. Many other species were also killed including some clams and other benthic invertebrates. Other sublethal effects of oil on benthic invertebrates include developmental problems such as slow growth, differential growth of body parts (deformity), changes in molting times, and occasional anomalies in development of organs. Reproductive effects such as smaller brood sizes and premature release of eggs, reduced feeding, and increased respiration have also been noted in tests. Benthic infauna will sometimes leave their burrows, exposing themselves to predators. Starfish will often retract their tube feet and lose their hold as a result.

Benthic invertebrates can take up hydrocarbons by feeding on contaminated material, breathing in contaminated water, and through direct absorption from sediments or water. Most invertebrates depurate hydrocarbons when the water and sediment return to a clean state or if placed in a clean environment. In severe oiling, however, depuration can take months. Sessile (or immobile) species are obviously at a disadvantage and may perish from prolonged exposure to contaminated sediments. Generally, however, all benthic species are affected by a short-term dose of the hydrocarbons in oil.

Epontic Organisms

Epontic organisms are microscopic plants and animals that live under ice. Many of these are similar to plankton and have similar responses and sensitivities to oil. Epontic organisms are much more vulnerable than

PHOTO 13.6
Oiled bird eggs in a nest. These eggs will be destroyed as a result of this oiling.

plankton, however, because oil remains directly under the ice, where these organisms live. Contact with oil causes death. The community may also be slow to recover because the oil can remain under the ice for a season or more, depending on the geographic location. Because the major limitation to growth for these organisms is the lack of room under the ice as well as low light and temperature levels, the dead organisms are not quickly replaced.

Marine Mammals

The effects of oil spills on marine and other aquatic mammals vary with species. Seals, sea lions, walruses, whales, dolphins, and porpoises are discussed here, as well as the effects on polar bears and otters. Although these two species are not actually marine mammals, they spend much of their time in or near the water. All of these animals are highly visible and cause much public concern when oiled.

Seals, sea lions, and *walruses* are particularly vulnerable to oiling because they live on the shorelines of small islands, rocks, or remote coasts with few options for new territory. Despite this, only the young are killed by severe oiling. External oiling of young seals or sea lions generally causes death because their coats are not developed enough to provide insulation in an oiled state. Oil is often absorbed or ingested, and mothers may not feed their young when they are oiled. After a large oil spill in South America, about 10,000 baby seals perished when the beaches of their island were contaminated by oil. Not many adult seals perished at the same site, and those who did probably drowned. Most of the oiled young seals died by starvation after their mothers refused to feed them.

Older seals, sea lions, and walruses can take a large amount of oiling without causing death. If lightly oiled, adult seals survive and the oil is slowly lost. Oiling of both adult and young causes the fur to lose waterproofing and buoyancy. It is not known if seals or their relatives would avoid oil if they could, as this has not been observed at spill sites.

Brief exposure of seals, sea lions, and walruses to volatile oil causes eye irritation and longer exposure can cause more permanent eye damage. Several studies on ingested oil have shown that hydrocarbons accumulate in the blubber, liver, kidney, and other organs, although the levels diminish within a few weeks. Long-term effects have not been observed and are difficult to measure because of the difficulty of approaching relatively healthy seals, sea lions, and walruses.

Whales, dolphins, and *porpoises* can be exposed to oil in the water column or on the surface when they come up to breathe. Despite this, deaths of these species have not been reported as a result of a spill. This is probably due to a number of factors. Oil does not adhere to the skins of these mammals and, as they are highly mobile, they are not exposed for a long period of time. Whales and dolphins have been observed to avoid oil spills and

PHOTO 13.7
A group of oiled brown pelicans awaits cleanup. (Photo from Flickr.com: http://www.flickr. com/photos/bpamerica/sets. Accessed April 24, 2012.)

contaminated waters. There is little information on the effect of ingested oil on whales and their kin, nor is there any evidence that hydrocarbons would be absorbed from water.

Polar bears spend much of their time near or in water, swimming between ice floes hunting seals. The potential for oiling is moderate. It was found that polar bears that are oiled ingest oil through grooming themselves, resulting in death or severe illness. Unfortunately, polar bears are attracted to oil, particularly lubricating oil, which they will actually drink. This generally causes temporary illness, but in the case of an oil spill, it could result in death. Few studies have been done of the sublethal effects of oil on polar bears as they are difficult to study.

Otters live on or near shorelines, and spend much of their time on the water or feeding on crustacea on the seafloor. Otters are usually oiled in any spill near their habitats and can die after only a 30% oiling. Oil adheres to the otters' fur causing heat loss, which is the most pronounced effect of oiling. Otters attempt to groom themselves after oiling and thus ingest oil, compounding their difficulty. As in the case of the polar bear, little is known about the effects of ingested oil. Some inflammation of the stomach and uptake of hydrocarbons have been observed. After light exposures, however, the animals appear to recover.

Oiled otters are often caught and taken to rehabilitation centers for cleaning by trained specialists. If caught and treated soon enough, some otters can be saved. Such rehabilitation is difficult and expensive, however, as animals may have to be kept for a month before release. In addition, many animals die after their release, possibly as a result of human handling.

Intertidal Fauna

Intertidal fauna include animals that live in the shoreline zone between the high and low tides. These organisms are the most vulnerable to oil spills because they and their habitat are frequently coated during oil spills. Typical fauna include the mobile crabs, snails, and shrimp; sessile (immobile)

PHOTO 13.8
A brown pelican is released after cleanup and rehabilitation. (Photo from Flickr.com: http://www.flickr.com/photos/bpamerica/sets. Accessed April 26, 2012.)

barnacles and mussels; and sedentary limpets, periwinkles, and tube worms. Heavy oiling will generally kill most species. The area does recolonize after the spill with the mobile species returning first, but recovery takes months and sometimes years. Recolonization by plants and sessile species is the major factor in site restoration.

As with other aquatic organisms, light oiling affects the immobile species most and most species will take up oil. Mussels and crabs in particular have been studied for their response to oil. Sublethal effects include reduced growth and reproduction rate, and accumulation of hydrocarbons. Both mussels and crabs will depurate or cleanse themselves of hydrocarbons when placed in clean water. Crabs also show premature or delayed molting. Mussels reduce production of attachment threads, often causing the creature to let go of its hold on its feeding surface. Other intertidal fauna show similar behavior as a result of light oiling.

Shoreline cleaning techniques have a strong effect on the recovery of an intertidal area. Very intrusive techniques such as washing with hot or high-pressure water can remove many of the food sources and thus delay recovery, despite removing all of the oil. Intertidal fauna are not highly affected by nonvolatile residual oil unless they are coated with it. Recovery is fastest in those areas where oil is removed rapidly after a spill using a nonintrusive technique such as cold water, low-pressure washing.

Marine Plants

Marine plants cover a wide spectrum of plant families and algae. Intertidal algae, macroalgae, and sea grasses are of particular interest during oil spills.

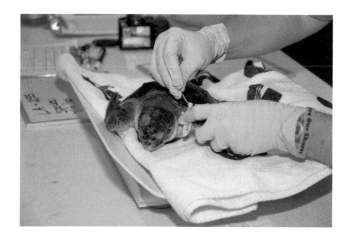

PHOTO 13.9
A sea turtle is tagged after its cleaning and rehabilitation. (Photo from the U.S. Coast Guard Web site: http://cgvi.uscg.mil.)

Intertidal algae are an important food source for much of the intertidal fauna and like the fauna can be severely affected by an oil spill. Although readily killed by even a moderate oil spill, intertidal algae are usually the first biota to recover after a spill. These algae grow on rock and sediment surfaces and will not recolonize if these surfaces are still heavily oiled with a light oil. Algae will reestablish on oil-coated rocks if the oil is weathered and no longer gives off volatile compounds. Like intertidal fauna, algae are also vulnerable to intrusive cleaning techniques such as washing with hot or high-pressure water. In fact, more algae are killed by these techniques than by oil. Sublethal effects include reduced reproduction and respiration rates, and changes in color.

Macroalgae include two common groups of plants in North America, *Fucus* and kelp, both of which include many species and subspecies. *Fucus*, which often inhabit the lower intertidal and subtidal zones, are not particularly susceptible to oiling because a mucous coating prevents the oil from adhering to the plant. A heavy oil will cover *Fucus*, however, and cause death or sublethal effects. Kelp generally lives in deeper water and is rarely coated with oil. Both *Fucus* and kelp will absorb hydrocarbons in the water column, but their effect, including death, depends on the length of time that the concentrations are present. A dose of a few hours will cause only slight and sublethal effects, whereas a moderate concentration over a few days could cause more serious damage and even death. Both plants will show sublethal effects of leaf loss, color changes, reproductive slowdown, reduced growth, and accumulation of hydrocarbons. Both plants will also slowly depurate or cleanse themselves of hydrocarbons in clean water. As these plants make up the habitat for complex ecosystems including many forms of animals and other algae, the entire ecosystem can be affected if they are damaged. Recovery for both types of plants and their habitats may take years.

Sea grasses generally inhabit the low intertidal and subtidal zones, and are extensive in any location around the world. Eelgrass, which is the common species in North America, is a vascular plant similar to most common water plants. Sea grasses are sensitive to hydrocarbon uptake and oiling. Because direct oiling rarely occurs, uptake of hydrocarbons from the water column is the main concern. Eelgrass is killed by moderate hydrocarbon concentrations in the water for a few hours or low concentrations for a few days. Eelgrass will show similar sublethal effects as kelp and *Fucus*, and will also depurate or cleanse itself of hydrocarbons. A bed of eelgrass killed by an oil spill may take several years to recover.

Special Ecosystems

Salt or *brackish marshes* are important ecosystems because they are the habitat of many birds and fish that feed on a wide variety of invertebrates including

PHOTO 13.10
A Northern Gannet is cleaned after a spill. (Photo from Flickr.com: http://www.flickr.com/photos/bpamerica/sets. Accessed April 26, 2012.)

crabs, snails, and worms. Some of these organisms burrow into the sediments providing a path for oil to penetrate if a spill occurs. They are also the nurseries for many land and seabirds and animals. Salt marshes are especially vulnerable to oil spills because they are flooded at high tide and their complex surface traps large quantities of oil. It is also difficult to get into a marsh to assess the damage and clean the oil.

Salt marshes are dominated by marsh grasses, the predominant one in temperate climates being *Spartina* and in the Arctic, *Puccinellia*, which has similar characteristics. The outer fringes of marshes are dominated by shrubs such as sedges. Marshes also export a large amount of plant detritus back to the sea, which contributes to the food chains of connecting water bodies.

The effects of oil on a marsh depend on the amount and type of oil. Light to moderate amounts of a weathered oil or an oil that does not penetrate significantly will not result in massive mortality and the marsh can recover in as little as 1 to 2 years. Heavy oiling by a light oil that penetrates the sediments can cause heavy mortality and the marsh can take up to 10 years to recover. Due to the dynamic, constantly changing nature of marshes, oil can be covered and retained in relatively unweathered conditions for decades. Massive oiling causes loss of the plant cover, which would also affect the animals and birds living in the marsh. As *Spartina* propagates from special root parts, any cleaning activity that destroys these will kill the plants. Marshes are very sensitive to physical disturbance and intrusive cleanup could easily cause more damage than the oil itself.

Arctic environments are often cited as a special case for oil spills, but in fact, extensive work on the toxicity and effects of oil have shown that Arctic species are about equally sensitive to oiling as their southern equivalents. The impact of an oil spill is increased, however, by the fact that the diversity of biota in the Arctic is very low and it takes longer to develop and grow. As oil takes longer to degrade and weather in the Arctic, toxic, volatile components are retained longer. For all these reasons, recovery from an oil spill is slower in an Arctic environment than in temperate and tropical zones.

Coral reefs occupy a large part of the seas in the tropics of the Pacific and the Caribbean. They are the most diverse and complex marine communities, supporting thousands of fish, algae, and invertebrate species. Studies and actual spills have shown that moderate concentrations of dissolved or dispersed hydrocarbons can kill both the coral and its occupants. Damage depends on the depth, with coral that is near the surface (down to about 6 m) being particularly vulnerable to oil. Many of the animals can repopulate the area rapidly, but since the coral is their primary support, full recovery depends largely on the recovery or recolonization of the coral. Once dead, the coral itself can be very slow to recover. Oil also has several sublethal effects on coral, such as slowed growth or respiration and unnatural coloration. Infant free-swimming coral are particularly sensitive to oil and have no membrane to prevent oil adsorption into their bodies. Even the smallest amount of dissolved oil is lethal to this life-form of coral.

PHOTO 13.11
A spill assessment worker walks through an oiled marsh.

Mangroves are trees that grow along much of the shorelines in the tropics. They provide the habitat for a wide diversity of other life. Mangroves are supported by a complex, interlaced system of prop roots. The base of the roots is in low-oxygen soil, and the trees take in air through breathing holes on the prop roots. If these are oiled and most of the breathing pores are plugged, the mangrove may die. The many other animals or birds supported by the trees are also at risk. It could take years to decades for mangroves to grow back in the oiled area. As with most plants, mangroves are subject to a number of sublethal effects including slower growth, loss of leaves, and changes in color.

Land

Unlike on water, oil spilled on land does not spread quickly and the effects remain local. Most types of oil will penetrate the soil and contaminate organisms in the soil. Diesel fuel was used at one time as a general vegetation killer. A full coating of fresh crude oil or diesel fuel will kill most plants and small trees on contact. Because of the low potential for affecting plants and less mobile animals directly on-site, however, the effects of oil on land environments are not as great a concern as in marine environments. Oil spills on land are discussed in Chapter 12.

Birds

Birds are the most visible biota affected by oil spills, especially in the aquatic environment. Oil contaminates feathers when the birds come into contact

PHOTO 13.12
Water birds after oil cleanup swim about an enclosed pool. (Photo courtesy of Environment Canada.)

with slicks on water or shorelines. For seabirds, this is particularly dangerous because when their feathers are oily, their insulation and buoyancy properties are decreased. Once oiled, a bird rapidly loses its body heat, especially at sea, possibly causing death. Oiled seabirds may stay on land where their temperature loss is not as great. In doing so, however, they are away from their source of food and may die of starvation.

Birds clean their plumage by preening and in doing so may ingest some of the oil. Birds may also ingest oil by eating oiled prey. Although ingestion of oil may cause death, it is more likely to cause sublethal effects such as gastrointestinal dysfunction, liver problems, pneumonia, and behavioral disorders.

Contaminated birds may transfer oil to their eggs or young. It has been found that only a few drops of fresh oil can kill the young in an egg. Even when birds ingest only a small amount of oil, they may stop laying eggs or the number of eggs may be reduced. A small amount of oil can also affect the hatchability of the eggs.

Shoreline dwellers and feeders, which include ducks, gannets, and cormorants, are among the most susceptible birds to oiling. Auks and ducks that spend much of their time on the water are also susceptible to oil spills at sea. These birds feed by diving through the surface. Endangered species and those concentrated in a few colonies are particularly vulnerable as a spill could threaten the entire species.

In many spills, cleaning stations are set up to rehabilitate birds. Although techniques have improved greatly in the past few years, success rates are still poor as it is very stressful for a wild bird to be captured and handled.

Less than half of the oiled birds that are cleaned and released actually survive. Only very sick birds can generally be captured and thus many of the birds brought to the treatment centers are often near death. Despite this, cleaning birds is easier than cleaning mammals and can reverse some of the effects of an oil spill.

Damage Assessment

Damage assessment is a formal, structured examination of an oiled environment to determine how many of each species was affected by the oil spill. The objectives are to quantify the damage to the environment as much as possible and assess the total effects of a particular spill. Data are used to develop long-term restoration or cleanup plans if necessary, to assess costs, and to provide a database of spill damage. Damage assessment involves a thorough reexamination of the site through counts of plants and animals and a comparison to the prespill condition. If information on the prespill condition is not available, the site is compared to a similar unoiled site nearby.

In the United States, damage assessment is becoming mandatory after oil spills and procedures have been developed to assess the costs of damages. A computer program has been developed to assist in performing these assessments. Damage assessment is very difficult, however, especially estimating the cost of specific damage to a specific resource.

Restoration

Site restoration arises from damage assessment. Sites are restored by replanting trees and vegetation, and recolonizing animals and birds at a site. Although this appears simple and beneficial, it is fraught with difficulties and can upset the ecological balance in some areas if not carried out carefully. It is difficult and sometimes impossible to recolonize or move certain species of animals. Furthermore, a damaged site will often require a succession of different plant and animal species before a balance is achieved. If not carefully planned and conducted, human intervention can upset this natural succession process.

Despite these difficulties, many badly oiled sites have been restored to almost their original state in several years. For example, a badly oiled marsh in New Jersey that scientists thought would be impossible to restore began to recover a year after marsh plants were transplanted and some native animals were returned to the site.

Net Environmental Benefits

Oil spill responders try to optimize net environmental benefits when considering how to deal with a spill. This simply means that the effects on the environment of whatever cleanup techniques are to be used are weighed against the damage to the site. In other words, the question is asked, Will the cleanup process itself possibly cause more damage to the site than the oil itself if it were left? Sometimes the decision is made not to clean if an assessment shows that the cleanup itself will be intrusive. In the same way, the effects of the various cleanup techniques are also assessed and the least intrusive technique is chosen for a particular site.

Glossary

Absorption: This is a process whereby one substance penetrates the interior of another substance. In the case of oil spill cleanup, this process takes place in the form of uptake of oil by capillaries within certain sorbent materials. (See also **Capillary action.**)

Adsorption: This is the process whereby one substance is attracted to and adheres to the surface of another substance without actually penetrating its internal structure. This is the most common uptake process for sorbents as described in this book.

Air or water streams: This is a method of oil containment in which the force of air or water directed as a stream can be used to divert or contain an oil slick. The method can be used to flush oil from beneath docks or to direct oil to a skimmer.

Alcohols: These are a class of organic chemical compounds characterized by the presence of the hydroxyl (OH oxygen-hydrogen) group attached to a carbon atom. Alcohols are important solvents and are used to a certain extent in preparing chemical dispersants. (See also **Glycols.**)

Alkanes: These are a class of hydrocarbons (compounds of hydrogen and carbon) that make up the primary part of the saturate group of components in oil. They are characterized by branched or unbranched chains of carbon atoms with attached hydrogen atoms. Alkanes all have the general formula C_nH_{2n+2} and contain no carbon-carbon double bonds, that is, they are "saturated" with hydrogen. Alkanes are also called paraffins and are a major constituent of natural gas and petroleum. Alkanes containing less than five carbon atoms per molecule ("n" in above formula is less than five) are usually gases at room temperature, for example, methane; those with 5 to 15 carbons are usually liquids, and straight chain alkanes with more than 15 carbons are solids. At low concentrations, alkanes with low carbon numbers may produce anesthesia and narcosis (stupor, slowed activity) and at high concentrations can cause cell damage and death in a variety of organisms. Alkanes with a higher number of carbon atoms are not generally toxic but have been shown to interfere with normal metabolic processes and communication in some species. (For other common hydrocarbons, see also **Alkenes, Aromatics, Naphthenes, Olefins, Paraffin, Saturate group.**)

Alkenes: These are a class of straight or branched chain hydrocarbons similar to alkanes but characterized by the presence of carbon atoms united by double bonds. Alkenes are also called olefins and all have the general formula C_nH_{2n}. Alkenes containing two to four carbon atoms are gases at room temperature, while those containing five or more carbon atoms are usually liquids. Alkenes are not found in crude oils but are often formed in large quantities during the cracking (breaking down of large molecules) of crude oils and are common in many refined petroleum products such as gasolines. These hydrocarbons are generally more toxic than alkanes but less toxic than aromatics. (For other classes of hydrocarbons, see also **Alkanes, Aromatics, Naphthenes, Olefins, Saturate group.**)

Ambient: This is used to refer to local or surrounding conditions, primarily climatic conditions, at some point in time, for example, ambient temperature.

Anaerobic: This is a term used to describe a situation or an area characterized by the lack of oxygen. The term can also be used in reference to organisms such as some bacteria that can survive and grow in the absence of gaseous or dissolved oxygen. For example, many marine sediments are anaerobic below a depth of a few centimeters from the surface. Oil deposited in such areas degrades slowly and is primarily associated with anaerobic types of microorganisms.

API gravity: This is a scale developed by the American Petroleum Institute (API) to designate an oil's specific gravity or the ratio of the weights of equal volumes of oil and pure water. API gravity is dependent on temperature and barometric pressure and is therefore generally measured at 16°C and one atmosphere pressure. Water with a specific gravity of 1.0 has an API gravity of 10°. A light crude oil may have an API gravity of 40°C. Oils with low specific gravities have high API gravities and vice versa. API gravity can be calculated from specific gravity using the following formula:

$$\text{API gravity} = (141.5/\text{Specific gravity @ } 15.5°C) - 131.5$$

(See also **Specific gravity.**)

Aromatics: This is a class of hydrocarbons considered to be the most immediately toxic hydrocarbons found in oil and that are present in virtually all crude oils and petroleum products. Many aromatics are soluble in water to some extent, thereby increasing their danger to aquatic organisms. Certain aromatics are considered long-term poisons and often produce carcinogenic effects. Aromatics are characterized by rings containing six carbon atoms. Most aromatics are derived from benzene, which is the simplest aromatic. In benzene, three double carbon-to-carbon bonds float around the ring containing six carbon atoms, which makes the benzene rings very stable and hence persistent in the environment. (For other classes of hydrocarbons, see also **Alkanes, Alkenes, Naphthenes, Olefins, Saturate group.**)

Asphalt: This is a black or brown hydrocarbon material made up primarily of the larger polar compounds called asphaltenes and aromatics.

Asphalt ranges in consistency from a heavy liquid to a solid. The most common source of asphalt is the residue left after the fractional distillation of crude oils. Asphalt is primarily used for surfacing roads. (See **Asphaltenes, Polar compounds.**)

Asphaltenes: These are the larger polar compounds found in oil, so named because they make up the largest percentage of the asphalt used to pave roads. Asphaltenes often have very large molecules (or a high molecular weight) and because of this, they biodegrade very slowly. This explains the durability of asphalt roof shingles and pavement. If there are enough asphaltenes in an oil, they greatly affect how the oil behaves when spilled. (See also **Asphalt, Polar compounds.**)

Backshore: This is the area of the shoreline above the high-tide mark. As the backshore is inundated with water only during exceptionally strong storms or abnormally high tides accompanied by high winds, it does not support characteristic intertidal flora and fauna. Granular materials to replace oil-contaminated beach material excavated during shoreline cleanup programs are frequently taken from backshore areas.

Barrel: This is a unit of liquid (volumetric) measure for petroleum and petroleum products, equal to 35 imperial gallons, 42 U.S. gallons, or approximately 160 liters (L). This measure is used extensively by the petroleum industry.

Benthos: This refers to the environment on the seafloor and includes plankton and fish.

Biodegradation: This refers to the degradation of substances resulting from their use as food energy sources by certain microorganisms such as bacteria, fungi, and yeasts. The process of oil degradation is extremely slow and is greatly limited by temperature, nutrients, and the availability of oxygen. Although more than 200 species of microorganisms have the ability to utilize hydrocarbons as an energy source, no single species can degrade more than about 10 of the many compounds normally found in oil. (See also **Weathering.**)

Biodegradation agents: These agents are used primarily on shorelines or land to accelerate the biodegradation of oil in the environment. They include bioenhancement agents that contain fertilizers or other materials to enhance the activity of hydrocarbon-degrading organisms, bioaugmentation agents that contain microbes to degrade oil, and combinations of these two.

Biological productivity: This is a measure of the biological activity of a population, community, or ecosystem, and is usually expressed as the quantity of carbon stored in tissue per unit of time. Certain environments are characterized by higher biological productivity than others. For example, marshes and estuaries are generally more productive than offshore marine waters. The biological productivity of an area is an important consideration when preparing contingency plans and assigning priorities for oil spill cleanup.

Boiling point: This is the temperature at which a liquid begins to boil. Specifically, it is the temperature at which the vapor pressure of a liquid is equal to the atmospheric or external pressure. The boiling point of crude oils and petroleum products may vary from 30°C to 550°C but is of little practical significance in terms of oil spill cleanup.

Boom failure: This refers to the failure of a containment boom to contain oil due to excessive winds, waves, or currents or improper deployment. Boom failure may be manifested in oil underflow, oil splashover, submergence or planing of the boom, or structural breakage. (See also **Critical velocity, Entrainment failure**.)

Brine channels: These are small passages in the lower surface of first-year sea ice, which are formed by the exclusion of saline water or salts during rapid freezing. Researchers have found that oil under first-year ice will migrate through the brine channels when the ice begins to melt in the spring. (See also **First-year ice**.)

Bubble barrier: This is a method for containing oil consisting of an underwater air delivery system that creates a curtain of rising bubbles that deflects the oil. The system has been used with some success in relatively calm areas, such as harbors. The system requires considerable maintenance when the submerged perforated pipes used to produce the bubble curtain become covered with redistributed bottom silt.

Bulk carrier: This is an oceangoing vessel designed to transport large quantities of a single product such as grain, ore, or coal.

Bunker B: This is a relatively viscous fuel oil (No. 5 fuel) used primarily as a fuel for marine and industrial boilers.

Bunker C: This is a very viscous fuel oil (No. 6 fuel) used as a fuel for marine and industrial boilers.

Burn efficiency: When carrying out in-situ burning of an oil spill, this is the percentage of the oil removed from the water by burning. It is the amount of oil before burning, less the amount remaining as a residue, divided by the initial amount of oil.

Burn rate: When carrying out in-situ burning of oil spills, this is the rate at which oil is burned within a given area or the rate at which the thickness of the oil diminishes. In most situations, the burn rate is approximately 2 to 4 mm/minute. (See also **In-situ burning**.)

Capillary action: This is the process whereby the force of attraction between a solid and a liquid causes the liquid to be drawn into the porous internal structure of the solid. (See also **Absorption**.)

Carbon number: This is the number of carbon atoms present in a single molecule of a given hydrocarbon. The physical and chemical properties of hydrocarbons tend to vary with the number of carbon atoms and these properties are frequently described in terms of range of carbon numbers for specific classes of hydrocarbons. For example, alkanes with carbon numbers from 1 to 4 are gaseous at ordinary temperatures and pressures.

Catalyst: This is a substance added to a reacting system, for example, a chemical reaction, which alters the rate of the reaction without itself being consumed. Most catalysts are used to increase the rate of a reaction. For example, the metal vanadium is often present in trace amounts in crude oils and acts as a catalyst to accelerate the rate of chemical oxidation of certain hydrocarbons as the oil weathers. Catalysts, such as silica and alumina, are also used during the refining of petroleum to increase the rate at which large hydrocarbon molecules are split into smaller ones, a process referred to as **catalytic cracking**.

Chemical barrier: This consists of chemicals that act as surface tension modifiers to inhibit the spread of an oil slick on water. When placed on the water surface next to an oil film, these chemicals push away the oil as a result of their surface tension. Chemical barriers work only with fresh oils, however, and their effect lasts only a few hours. (See also **Surface tension**.)

Chemical dispersion: In relation to oil spills, this term refers to the creation of oil-in-water emulsions by the use of chemical dispersants made for this purpose. (See also **Dispersants or chemical dispersants**.)

Chocolate mousse: This is the term used to describe a water-in-oil emulsion consisting of 50% to 80% water. These emulsions are sometimes stable and range in consistency from greaselike to solid. They are only formed with a relatively viscous oil with extensive asphaltene content in the presence of considerable wave action. (See also **Emulsification, Water-in-oil emulsion**.)

Containment: This is the process of preventing oil from spreading beyond the area where it has been spilled in order to minimize pollution and facilitate recovery.

Containment boom: This is a floating mechanical structure that extends above and below the water surface and is designed to stop or divert the spread or movement of an oil slick on the water. Booms consist of floats, a freeboard member to prevent oil from flowing over the top of the boom, a skirt below the water surface to prevent oil from being swept under the boom, and one or more tension members to support the entire boom. Booms are an integral part of virtually all cleanup programs after oil spills on water. (See also **Boom failure, Critical velocity, Freeboard**.)

Contingency plan: This is an action plan prepared in anticipation of an oil spill. This plan usually consists of guidelines developed for a specific industrial facility or an entire region to increase the effectiveness, efficiency, and speed of cleanup operations in the event of an oil spill and at the same time protect areas of biological, social, and economic importance.

Countermeasure: This is a method used or an action taken to prevent or control pollution by oil spills.

Critical velocity: This is the lowest speed or velocity of the water current that will cause loss of oil under the skirt of a containment boom. Critical velocity varies with specific gravity, viscosity, and thickness of the oil slick contained by the boom, and the depth of the skirt and position of the boom in relation to the direction of the current. For most oils, when the boom is at right angles to the current, the critical velocity is about 0.5 m/sec (1 knot). (See also **Boom failure.**)

Crude oils: This refers to petroleum in its natural form before it is subjected to any refining process such as fractional distillation or catalytic cracking. The main elements in crude oils are hydrogen and carbon as they are composed of mixtures of hydrocarbon compounds. Crude oils also contain varying amounts of sulfur, nitrogen, oxygen, and sometimes mineral salts, as well as trace metals such as nickel, vanadium, and chromium.

Density: This is the mass or weight of a given volume of oil and is typically expressed in grams per cubic centimeter (g/cm^3). The petroleum industry defines heavy or light crude oils in terms of this property. Density also indicates whether a particular oil will float on water. The density of oil increases with time as the light fractions evaporate.

Detritus: This is loose material resulting from rock disintegration or abrasion. This can also refer to suspended material in the water column including fragments of decomposing flora and fauna, and fecal pellets produced by zooplankton and associated bacterial communities.

Dispersants or chemical dispersants: These are chemicals that reduce the surface tension between water and a hydrophobic substance such as oil. In the case of an oil spill, disperants thereby facilitate the breakup and dispersal of an oil slick throughout the water column in the form of an oil-in-water emulsion. Chemical dispersants can only be used in areas where biological damage will not occur and must be approved for use by government regulatory agencies. (See also **Chemical dispersion.**)

Dispersion: This is the distribution of spilled oil into the upper layers of the water column either by natural wave action, by applying chemical dispersants, or by using one of various hydraulic dispersion techniques. (See also **Chemical dispersion, Weathering.**)

Dissolution: This is the act or process of dissolving one substance in another. Specifically, it is a process that contributes to the weathering of spilled oil whereby certain "slightly" soluble hydrocarbons and various mineral salts present in the oil are dissolved in the surrounding water. (See also **Solubility, Weathering.**)

Distillation fractions: These represent the fraction (generally measured by volume) of an oil that is boiled off at a given temperature. For example, while 70% of gasoline will boil off at 100°C, only about 5% of a crude oil will boil off at that temperature, and an even smaller amount of a typical Bunker C oil. The distillation fractions of an oil correlate strongly to the composition of the oil as well as to other physical properties of the oil.

Elevating skimmers: A type of mechanical skimmer designed to remove oil from the water surface using conveyors to lift oil into a recovery area in an operation similar to removing water from a floor with a squeegee.

Emulsification: This is the process whereby one liquid is dispersed into another liquid in the form of small droplets. In the case of oil, the emulsion can be either oil-in-water or water-in-oil. Both types of emulsions are formed as a result of wave action, although water-in-oil emulsions are more stable and create special cleanup problems. (See also **Chocolate mousse, Oil-in-water emulsion, Water-in-oil emulsion.**)

Emulsion breakers and inhibitors: These are chemical agents used to prevent the formation of water-in-oil emulsions or to cause such emulsions to revert to oil and water. Several formulations can perform both functions.

Entrained water: Oil of higher viscosity can uptake water that is not stabilized by asphaltenes and resins inside the oil. This water settles out of the oil once the oil is out of seas with high waves or turbulence.

Entrainment failure: This is a type of boom failure resulting from excessive current speed or velocity. The head wave formed upstream of the oil mass contained within a boom becomes unstable and oil droplets are torn off and become entrained or drawn into the flow of water beneath the boom. (See also **Boom failure, Critical velocity, Head wave.**)

Environmental sensitivity: This term is used to describe the susceptibility of a local environment or area to any disturbance that might decrease its stability or result in either short- or long-term adverse effects. Environmental sensitivity generally includes physical and biological factors.

Epontic organisms: These are microscopic plants and animals that live under ice. Many are similar to plankton although they are much more vulnerable to oil because oil remains directly under the ice where these organisms live. Contact with oil causes death. (See **Benthos, Plankton.**)

Estuary: This is a partly enclosed coastal body of water in which freshwater, usually originating from a river, is mixed with and diluted by seawater. Estuaries are generally considered more biologically productive than either adjacent marine or freshwater environments, and are important areas in the life history of many fish and wildlife resources.

Evaporation: This is the process whereby any substance is converted from a liquid state to become part of the surrounding atmosphere in the form of a vapor. In the case of oil, the rate of evaporation depends primarily on the volatility of various hydrocarbon constituents and temperature. Evaporation is the most important process in the weathering of most oils. (See also **Weathering**.)

Fauna: This term refers to animals in general or animal life as distinguished from plant life (flora). Fauna usually refers to all the animal life characteristic of or inhabiting a particular region or locality.

Fire-resistant booms: These are floating containment structures constructed to withstand high temperatures and heat fluxes, used when burning oil on water. These booms restrict the spreading and movement of oil slicks while increasing the thickness of the slick so the oil will ignite and continue to burn.

First-year ice: This is ice formed during the winter of a given year in the Arctic regions that does not contain residual or polar pack ice from previous years. First-year ice is characterized by the presence of brine channels and is relatively porous compared to pack ice. (See also **Brine channels**.)

Flash point: The flash point of an oil is the temperature at which the liquid gives off sufficient vapors to ignite when exposed to an ignition source such as an open flame. A liquid is considered to be flammable if its flash point is less than 60°C. (See also **Boiling point**.)

Flora: This term refers to plants and bacteria in general or plant and bacterial life as distinguished from fauna (animal life). Flora usually refers to all the plant life inhabiting or characteristic of a particular region or locality.

Fractional distillation: This is the separation of a mixture of liquids such as crude oils into components with different boiling points. Fractional distillation is the primary process in the refining of crude oils.

Freeboard: This is the part of a floating containment boom that is designed to prevent waves from washing oil over the top of the boom. The term freeboard is also used to describe the distance from the water surface to the top of the boom. Freeboard is generally also applied to the distance from the deck to the waterline of a vessel such as a ship or barge. (See also **Containment boom**.)

Freighter: This is a vessel or aircraft used primarily for carrying freight or goods.

Fuel oils: These are refined petroleum products with specific gravities of 0.85 to 0.98 and flash points greater than 55°C. This group of products includes furnace, auto diesel, and stove fuels (No. 2 fuels), industrial heating fuels (No. 4 fuels), and various bunker fuels (No. 5 and No. 6 fuel oils).

Gasolines: These are a mixture of volatile, flammable liquid hydrocarbons used primarily for internal combustion engines and characterized by a flash point of approximately −40°C and a specific gravity of 0.65 to 0.75.

Glycols: These are any of a class of organic compounds belonging to the alcohol family but with two hydroxyl groups (OH: oxygen-hydrogen) attached to different carbon atoms. The simplest glycol is ethylene glycol, a compound used extensively for automobile antifreeze. Other glycols are an ingredient in some chemical dispersants.

Groundfish: These are species of fish normally found close to the sea bottom throughout the adult phase of their life history. Groundfish feed extensively on bottom flora and fauna, and include species such as cod, halibut, and turbot.

Groundwater: This is water present below the soil surface and occupying voids in the porous subsoil, specifically the porous layer that is completely saturated with water. The upper surface of the groundwater is referred to as the water table. Contamination of groundwater is a major concern when oil is spilled on land as groundwater supplies springs and wells, and passes into surface water in many areas.

Head wave: This is an area of oil concentration that occurs in front of and at some distance from containment booms. This area of oil thickening is important to the positioning of mechanical recovery devices such as skimmers and is the region where the boom failure phenomenon known as entrainment is initiated when current flow exceeds critical velocity. (See also **Boom failure, Critical velocity, Entrainment failure**.)

Hydraulic dispersion: This is one of various shoreline cleanup techniques that uses a water stream at either low or high pressure to remove stranded oil. These techniques are most suited to removing oil from coarse sediments, rocks, and man-made structures, although care must be taken to avoid damage to intertidal flora and fauna.

Hydrocarbons: These are organic chemical compounds composed only of the elements carbon and hydrogen. Hydrocarbons are the principal constituents of crude oils, natural gas, and refined petroleum products, and include four major classes of compounds (alkanes, alkenes, naphthenes, and aromatics) each with characteristic structural arrangements of hydrogen and carbon atoms, as well as different physical and chemical properties. (See also **Alkanes, Alkenes, Aromatics, Naphthenes, Olefins, Paraffin, Saturate group**.)

Hydrophobic agent: This is a chemical or material that has the ability to resist wetting by water. Hydrophobic agents are occasionally used to treat synthetic sorbents to decrease the amount of water absorbed and hence increase the volume of oil they can absorb before becoming saturated. (See also **Synthetic organic sorbents**.)

Improvised booms: These are booms constructed from readily available materials such as railroad ties and logs. Improvised booms can be used as temporary containment structures until more suitable commercial booms arrive at the spill site. They can also be used in conjunction with commercial containment booms to divert oil into areas where the commercial booms are positioned.

In-situ burning: This is an oil spill cleanup technique or countermeasure that involves controlled burning of oil directly where it is spilled. It does not include burning oil or oiled debris in an incinerator.

Interfacial tension: This is the force of attraction or repulsion between the surface molecules of oil and water. It is sometimes referred to as **surface tension**.

Intertidal fauna: This includes animals that live in the shoreline zone between the high and low tide marks. Typical fauna include the mobile species of crabs, snails, and shrimp; and immobile species of barnacles, mussels, limpets, and periwinkles. These organisms are the most vulnerable to oil spills because they and their habitat are frequently coated during oil spills. Heavy oiling usually kills them.

Intertidal zone: This is the portion of the shoreline located between the high and low tide marks. This area is covered by water at some time during the daily tidal cycles. The size of the intertidal zone varies with the tidal characteristics of a given region as well as the steepness of the shoreline. In general, steep, rocky shorelines have smaller intertidal zones than gradually sloping mixed sediment beaches. Since all portions of the intertidal zone are without water cover at some time during each day, flora and fauna capable of withstanding desiccation are found throughout the zone. Depending on the level of the tide when oil reaches the shoreline, all or a portion of the intertidal zone can be affected by the incoming slick. The most suitable cleanup methods are dictated by the physical characteristics of the intertidal zone, as well as the sensitivity of the affected shoreline. After an oil spill, the intertidal zone is sometimes left to cleanse itself naturally.

Jet fuel: This is a kerosene or kerosene-based fuel used to power jet aircraft combustion engines. (See also **Kerosene**.)

Kerosene: A flammable oil characterized by a relatively low viscosity, specific gravity of approximately 0.8, and flash point close to 55°C. Kerosene lies between the gasolines and fuel oils in terms of major physical properties and is separated from these products during the fractional distillation of crude oils. Kerosene is used for wick lamps, domestic heaters and furnaces, fuel or fuel components for jet aircraft engines, and thinner in paints and insecticide emulsions. (See also **Jet fuel**.)

LC_{50}: This is a measure of the toxicity of a substance in terms of its median lethal concentration, that is, the concentration of material in water that is estimated to be lethal to 50% of a test population within a specified period of exposure. The duration of exposure must be specified, for example, a 96-hour LC_{50}.

Light ends: This is a term used to describe the low molecular weight, volatile hydrocarbons in crude oil and petroleum products. The light ends are the first compounds recovered from crude oil during the fractional distillation process and are the first fractions of spilled oil to be lost through evaporation.

Lubricating oils: These oils are used to reduce friction and wear between solid surfaces such as moving machine parts and components of internal combustion engines. Petroleum-based lubricating oils and greases are refined from crude oil through a variety of processes including vacuum distillation, extraction of specific products with solvents (solvent extraction), removal of waxes with solvents, and treatment with hydrogen in the presence of a catalyst. Most lubricating oils and greases derived from crude oil contain a large proportion of alkanes and naphthenes with high carbon numbers that provide the most favorable lubricating characteristics. The viscosity of a lubricating oil is its most important characteristic since it determines the amount of friction that will be encountered between sliding surfaces and whether a thick enough film can be built up to avoid wear from solid-to-solid contact.

Manual recovery: This is a term used to describe the recovery of oil from contaminated areas by cleanup crews with the use of buckets, shovels, and similar equipment. Manual recovery is extremely labor-intensive but plays a role in many oil spill cleanup programs.

Metric ton (tonne): This is a unit of mass and weight equal to 1000 kilograms or 2205 pounds avoirdupois. In Canada, the metric ton is the most widely used measure of oil quantity by weight. There are approximately 7 to 9 barrels (250 to 350 gallons) of oil per metric ton, depending on the specific gravity of the crude oil or petroleum product.

Microorganisms: These are microscopic- or ultramicroscopic-sized plant or animal life that is not visible to the human eye without the aid of a microscope. Microorganisms are present in the air, water, and soil, and generally include the bacteria, yeasts, and fungi. Some microorganisms are capable of metabolizing hydrocarbons and play a role in the natural degradation of spilled oil.

Mineral-based sorbent: These are any of a number of inorganic, mineral-based substances used to recover oil because of their adsorptive or absorptive capacities. Mineral-based sorbents include materials such as vermiculite, perlite, or volcanic ash and recover from 4 to 8 times their weight in oil. (See also **Natural organic sorbents, Synthetic organic sorbents.**)

Mineral spirits: These are flammable petroleum distillates that boil at temperatures lower than kerosene and are used as solvents and thinners, especially in paints and varnishes. Mineral spirits are the common term for some naphthas. Mineral spirits were used extensively in chemical dispersants made before 1970 but are no longer used due to their toxicity. (See also **Naphthas.**)

Molecular weight: This is the total mass of any group of atoms that are bound together to act as a single unit or molecule.

Naphthas: These are any of various volatile and often flammable liquid hydrocarbon mixtures used primarily as solvents and diluents. Naphtha consists mainly of hydrocarbons that have a higher boiling point than gasolines and lower boiling point than kerosene. Naphtha was a principle ingredient in chemical dispersants used before 1970. (See also **Mineral spirits.**)

Naphthenes: This is a class of hydrocarbons with physical and chemical properties similar to alkanes but characterized by the presence of simple closed rings. Like alkanes, naphthenes are also saturated, that is, they contain no carbon-carbon double bonds, and have the general formula, C_nH_{2n}. Naphthenes are found in both crude oils and refined petroleum products. They are insoluble in water and generally boil at $10°C$ to $20°C$ higher than alkanes with a corresponding number of carbons. (For other classes of hydrocarbons, see also **Alkanes, Alkenes, Aromatics, Saturate group.**)

Natural organic sorbents: These are natural materials such as peat moss, straw, and sawdust that can be used to recover spilled oil. Natural sorbents generally absorb 3 to 6 times their weight in oil by virtue of the criss-cross arrangement of fibers within the material. All natural sorbents absorb water as well as oil, however, and may sink when saturated with water. Indiscriminate use of natural sorbents can add to the problems of oil spill cleanup. Synthetic sorbents are usually favored due to their greater capacity for oil and relative ease of recovery. (See also **Mineral-based sorbents, Synthetic organic sorbents.**)

Oil slug: This refers to a downward moving oil mass that often results when oil is spilled on relatively porous soil. The slug-like shape results from the tendency of the descending oil mass to leave behind a funnel of soil that is partially saturated with oil.

Oil spill cooperative: These organizations are formed by oil companies operating in a given area for the purpose of pooling equipment, personnel, training, and expertise to combat oil spills.

Oil-in-water emulsion: This emulsion consists of oil droplets dispersed in surrounding water and is formed as a result of wave action or by a chemical dispersant. This is typically called a dispersion. Oil-in-water emulsions show a tendency to coalesce and reform an oil slick when the water becomes calm, although the presence of surface-active agents in the oil or artificially added in the form of chemical dispersants increases the persistence of this type of emulsion. Natural dispersion of large quantities of oil can follow the formation of oil-in-water emulsions and weathering processes such as dissolution, oxidation, and biodegradation may be accelerated due to the large increase in the surface area of the oil relative to its volume. (See also **Emulsification, Water-in-oil emulsion.**)

Olefins: This is a group of unsaturated hydrocarbon compounds that contain less hydrogen atoms than the maximum possible. Olefins have at least one double carbon-to-carbon bond that displaces two hydrogen atoms. Significant amounts of olefins are found only in refined products. (See also **Alkenes.**)

Oleophilic surface skimmers: These skimmers, which are also called sorbent surface skimmers, use a surface to which oil can adhere to remove oil from the water surface. This surface can be in the form of a disc, drum, belt, brush, or rope that is moved through the oil on the top of the water. The oil is removed with a wiper blade or pressure roller and deposited into an onboard container or pumped directly to a storage facility on a barge or on shore.

Oxidation or atmospheric oxidation: This is the chemical combination of compounds such as hydrocarbons with oxygen. Oxidation is a process that contributes to the weathering of oil. Compared to other weathering processes, however, oxidation is slow since the reaction occurs primarily at the surface and only a limited amount of oxygen penetrates the slick or surface oil. (See also **Weathering.**)

Oxygenated compounds: These are hydrocarbon compounds containing oxygen. They may be the result of incomplete combustion.

Paraffin: This is a waxy substance obtained from the distillation of crude oils and often contained in the crude oils. Paraffin is a complex mixture of alkanes with higher numbers of carbon that is resistant to water and water vapor and is chemically inert. The term is sometimes used to refer to alkanes as a class of compounds. (See also **Alkanes, Waxes.**)

Photooxidation: This occurs when the sun's action on an oil slick causes oxygen and carbons to combine to form new products that may be resins. The resins may be somewhat soluble and dissolve into the water, or they may cause water-in-oil emulsions to form. Some oils are more susceptible to photooxidation than others. In general, it is not an important process in changing an oil's properties after a spill. (See also **Resins, Weathering.**)

Phytoremediation: This is the use of plants and their associated microorganisms to degrade, contain, or render harmless contaminants in soil or groundwater.

PM-10: This is the particulate matter consisting of small respirable particles with a diameter of 10 μm (micrometers or microns) or less. PM-10 was once the standard for particulate concentration size, but has now been replaced by PM-2.5. For monitoring of particulate matter in the smoke plume from oil fires, it is generally accepted that the concentration of PM-10 particles should be less than 150 μg/m^3 for a 24-hour period. Particulate matter is the main public health concern when oil or petroleum products are burned. (See also **PM-2.5.**)

PM-2.5: This is the particulate matter consisting of small respirable particles with a diameter of 2.5 µm (micrometers or microns) or less. It has been found that 2.5 micrometers is a critical size below which human lungs are affected. For monitoring of particulate matter in the smoke plume from oil fires, it is generally accepted that the concentration of PM-2.5 particles should be less than 35 µg/m³ for a 24-hour period. Particulate matter is the main public health concern when oil or petroleum products are burned.

Polar compounds: These are hydrocarbon structures found in oil that have a significant molecular charge as a result of bonding with compounds such as sulfur, nitrogen, or oxygen. The "polarity" or charge carried by the molecule results in behavior that is different from that of unpolarized compounds under some circumstances. In the petroleum industry, the smallest polar compounds are called resins, which are largely responsible for oil adhesion. The larger polar compounds are called asphaltenes because they often make up the largest percentage of the asphalt commonly used in road construction. (See also **Asphaltenes, Resins.**)

Polyaromatic hydrocarbons (PAHs): These are the most common smaller and more volatile compounds found in oil. They contain multiple benzene rings. Crude oils and residual oils contain varying amounts of these compounds, some of which may be toxic to humans and aquatic life.

Polyethylene: This is a polymer (substance composed of very large molecules that are multiples of simpler chemical units) of the alkene ethylene. Polyethylene is highly resistant to chemicals, and has low water absorption and good insulating properties and can be manufactured in a number of forms. Polyethylene also has high oleophilic properties and has been used successfully as a sorbent for cleaning up oil spills. (See also **Alkenes, Oleophilic surface skimmers, Synthetic organic sorbents.**)

Polyurethane: This is any of a class of synthetic resinous, fibrous, or elastomeric compounds belonging to the family of organic polymers, consisting of large molecules formed by the chemical combination of successive smaller molecules into chains or networks. The best known polyurethanes are the flexible foams used as upholstery material and in mattresses and the rigid foams used as lightweight structural elements including cores for airplane wings. Polyurethane is also the most effective sorbent that can be used for oil spill cleanup and, unlike most synthetic sorbents, efficiently recovers oils with a wide range of viscosities. (See also **Synthetic organic sorbents.**)

Porosity: This is a measure of the space in rocks or soil that is not occupied by mineral matter. Porosity is defined as the percentage of total pore space in the total volume of rock, including all voids, whether interconnected or not. The porosity of a soil is determined by the mode of deposition, packing of grains, compaction, and grain shape and size. Porosity can also be used to refer to the voids in other materials such as sorbents.

Pour point: The pour point of an oil is the lowest temperature at which it will flow under specified conditions. The pour point of crude oils generally varies from –57°C to 32°C. Lighter oils with low viscosities have lower pour points. Pour point is defined as a minimum rate of movement and therefore is not a solidification point.

Recovery: In oil spill cleanup, this term applies to the entire process of physical removal of spilled oil from land, water, or shoreline environments, or any operation contributing to this process. General methods of recovering oil from water are the use of mechanical skimmers, sorbents, and manually by the cleanup crew. The main method of recovery of oil spilled on land or shorelines is by excavating contaminated materials.

Recovery agents or enhancers: Also called viscoelastic agents, these are chemical formulations intended to improve the recovery efficiency of oil spill skimmers or suction devices by increasing the adhesiveness of oil. One recovery enhancer consists of a polymer in the form of microsprings, or coiled macromolecules, which increase the adhesion of one portion of the oil to the other.

Remote sensing: This refers to the use of sensors other than human vision to detect or map oil spills. Remote sensing is usually carried out with instruments onboard aircraft or from satellite. The primary applications of remote sensing are locating an oil spill before its detection by any other means and monitoring the movement of an oil slick under adverse climatic conditions and during the night.

Residual oils: This is the oil remaining after fractional distillation during petroleum refining. This generally includes the bunker fuel oils.

Residue: This is the material, excluding airborne emissions, remaining on or below the surface after an in-situ burn takes place. It is largely unburned oil with some lighter or more volatile products removed.

Resins: These are the smallest polar compounds found in oil. They are largely responsible for oil adhesion. (See also **Photooxidation, Polar compounds.**)

Sanitary landfill site: This is an approved disposal area where materials including garbage, oil-contaminated debris, and highly weathered oil are spread in layers and covered with soil to a depth that will prevent disturbance or leaching of contaminants toward the surface. Sanitary landfill sites are located in areas where there is no potential for contamination of groundwater supplies. This method of disposal takes advantage of the ability of some bacteria and other microorganisms to biodegrade garbage and debris from oil spill cleanup operations.

Saturate group: This is a group of hydrocarbon components found in oils that consists primarily of alkanes, which are compounds of hydrogen and carbon with the maximum number of hydrogen atoms around each carbon. The term saturated is used because the carbons are "saturated" with hydrogen. The saturate group also includes cycloalkanes, which are compounds made up of the same carbon and hydrogen constituents but with the carbon atoms bonded to each other in rings or circles. Larger saturate compounds are often referred to as waxes. (See also **Alkanes, Hydrocarbons, Paraffins, Waxes.**)

Sedimentation: This is the process by which oil is deposited on the bottom of the sea or other water body. Most sedimentation seems to occur when oil droplets reach a higher density than water after interacting with mineral matter in the water column. Once oil is on the bottom, it is often covered with other sediment and degrades very slowly. (See also **Weathering.**)

Sediments: This is a general term used to describe material on the bottom and in suspension in water and the suspended material transported by a stream or river, the unconsolidated sand and gravel deposits of river valleys and coastlines, and materials deposited on the floor of lakes and oceans.

Sensitivity maps: These are maps used by the oil spill response team that designate areas of biological, social, and economic importance in a given region. These maps often prioritize sensitive areas so that in the event of an extensive spill, these areas can be protected or cleaned up first. Sensitivity maps usually contain other information useful to the response team such as the location of shoreline access areas, landing strips, roads, communities, and the composition and steepness of shoreline areas. These maps often form an integral part of the local or regional oil spill contingency plans.

Sheen: This is the common term used to describe a thin film of oil, usually less than 2 µm (0.002 mm) thick, on the water surface. (See also **Slick.**)

Shoreline sensitivity: This term refers to the susceptibility of a shoreline environment to any disturbance that might decrease its stability or result in short- or long-term adverse effects. Shorelines that are susceptible to damage from stranded oil are usually equally sensitive to cleanup activities, which may alter physical habitat or disturb associated flora and fauna. Marshes and lagoons are the most sensitive shoreline environments, while exposed coastlines subject to heavy wave action are generally the least affected by oil and cleanup activities.

Shoreline type: This is determined by the average slope or steepness and predominant substrate composition of the intertidal zone of a shoreline area. In any given region, shoreline type may be used to assess the type and abundance of intertidal flora and fauna, protection priorities, and the most suitable oil spill cleanup strategies.

Sinking agent: This is a material that is spread over the surface of an oil slick to adsorb oil and cause it to sink. Common sinking agents include treated sand, fly ash, and special types of clay. These materials are rarely used, however, because they provide a purely cosmetic approach to oil spill cleanup and may cause considerable damage to bottom-dwelling organisms. Use of sinking agents is generally forbidden.

Slick: This is the common term used to describe a film of oil resulting from a spill and varying from 2 μm to 2 cm. (See also **Sheen**.)

Solidifiers: These are a type of chemical spill-treating agent that are intended to change liquid oil to a solid compound that can be collected from the water surface with nets or mechanical means. They consist of cross-linking chemicals that couple two molecules or more to each other.

Solubility: This is the amount of a substance (solute) that will dissolve in a given amount of another substance (solvent). In terms of oil spill cleanup, it is most often the measure of how much oil will dissolve in the water column on a molecular basis. This is important in that the soluble fractions of the oil are sometimes toxic to aquatic life, especially at high concentrations. The solubility of oil in water is very low, generally less than 100 parts per million (ppm). (See also **Dissolution**.)

Sorbent: This is a substance that either adsorbs or absorbs another substance. Specifically, it is a natural organic, mineral-based, or synthetic organic material used to recover small amounts of oil that have been spilled on land or water surfaces or stranded on shorelines. (See also **Mineral-based sorbents, Natural organic sorbents, Synthetic organic sorbents**.)

Sorbent booms: These are specialized containment and recovery devices constructed of porous sorbent materials to absorb spilled oil while it is being contained. Sorbent booms and barriers are only used when the oil slick is relatively thin since their recovery efficiency rapidly decreases once the sorbent is saturated with oil. They are not absorbent enough to be used as a primary countermeasure technique for any significant amount of oil.

Sorbent surface skimmer: This is a mechanical skimmer that incorporates a rotating, sorbent surface (oleophilic) drum, disc, belt, or rope to which oil adheres as the surface is moved continuously through the slick.

Specific gravity: This is the ratio of the weight of a substance such as an oil to the weight of an equal volume of water. Buoyancy is intimately related to specific gravity; if a substance has a specific gravity less than that of a fluid, it will float on that fluid. The specific gravity of most crude oils and refined petroleum products is less than 1.0 and therefore these substances generally float on water. (See also **API gravity, Density**.)

Submersion skimmers: This mechanical skimmer incorporates a moving belt inclined at an angle to the water surface in such a way that oil in the path of the device is forced beneath the surface and subsequently rises (due to its buoyancy) into a collection well.

Substrate: This is material such as water, soils, and rocks, as well as other plants and animals that form the base of something. In biology, substrate refers to the base on which an organism lives.

Suction or vacuum skimmers: These skimmers use a vacuum or slight differential in pressure to remove oil from the water surface. They often consist of a small floating head, which is simply an enlargement of the end of a suction hose and a float, connected to an external source of vacuum, such as a vacuum truck.

Surface tension: This is the force of attraction or repulsion between the surface molecules of liquid. Usually the term "interfacial tension" is more correct. Surface tension affects the rate at which spilled oil will spread over a water surface or into the ground. Oils with low specific gravities are often characterized by low surface tensions and therefore have faster spreading rates. (See also **Interfacial tension.**)

Surface-washing agents: These are surfactant-containing agents designed to be applied to shorelines or structures to release oil from the surface.

Synthetic organic sorbents: These are one of several organic polymers, generally in the form of plastic foams or fibers, used to recover spilled oil. Synthetic organic sorbents have higher recovery capacities than either natural organic sorbents such as peat moss or mineral-based sorbents such as vermiculite. Many synthetic organic sorbents can be reused after the oil is squeezed out. Some common synthetic sorbents include polyurethane foam, polyethylene, and polypropylene. (See also **Hydrophobic agents, Mineral-based sorbents, Natural organic sorbents, Polyurethane.**)

Tanker: This is a cargo ship fitted with tanks for transporting liquid in bulk.

Tar: This is a black or brown hydrocarbon material that ranges in consistency from a heavy liquid to a solid. The most common source of tar is the residue left after fractional distillation of crude oil.

Tar balls or mats: These are compact, semisolid, or solid masses of highly weathered oil formed through the aggregation of viscous hydrocarbons with a high carbon number and debris in the water column. Tar balls are usually washed up on shorelines where they tend to resist further weathering.

Tension member: This is the part of a floating oil containment boom that carries the load placed on the barrier by wind, wave, and current forces. Tension members are commonly made of wire cable or chain due to their strength and stretch resistance.

Tide pools: These are permanent depressions in the substrate of intertidal zones that always contain water but are periodically flushed with successive incoming tides. Tide pools are often located near the high tide mark and contain abundant flora and fauna that can be adversely affected when spilled oil becomes stranded in these areas.

Total petroleum hydrocarbon (TPH): A simplified analysis of the amount of petroleum-origin material in water or soil. Older methods used solvent extraction and simple light measurement techniques. Modern methods use extraction and measurement by gas chromatography.

Toxicity: This is the capability of a poisonous compound or toxin to produce adverse effects in organisms. These effects include alteration of behavioral patterns or biological productivity, which is referred to as sublethal toxicity, or, in some cases, death or lethal or acute toxicity. The toxic capability of a compound is frequently measured by its "acute LC_{50}" with a standard test organism such as rainbow trout. This is the concentration that will result in death in 50% of the test organisms over a given time period, usually 96 hours. The most immediately toxic compounds in crude oils or refined petroleum products are the aromatics such as benzene. (See also **Aromatics, LC_{50}.**)

Ultraviolet radiation: This is the portion of the electromagnetic spectrum emitted by the sun, which is adjacent to the violet end of the visible light range. Often called "black light," ultraviolet light is invisible to the human eye, but when it falls on certain surfaces, it causes them to fluoresce or emit visible light. Ultraviolet light is responsible for the photooxidation of certain compounds including hydrocarbons although the process is limited in water, air, or soil by the low penetration ability of this short wavelength form of energy. (See also **Photooxidation.**)

Vapor pressure: This is a measure of how oil partitions between the liquid and gas phases, or how much vapor is in the space above a given amount of liquid oil at a given temperature.

Viscosity: This is the property of a fluid, either gas or liquid, by which it resists a change in shape or movement or flow. Viscosity denotes opposition to flow and may be thought of as an internal friction between the molecules in a fluid. For example, gasoline has a low viscosity and flows readily, whereas tar is very viscous and flows poorly. The viscosity of oil is largely determined by the amount of lighter and heavier fractions that it contains. The greater the percentage of light components such as saturates and the lesser the amount of asphaltenes, the lower the viscosity. Viscosity increases as oil weathers and as the temperature decreases. In terms of oil spill cleanup, viscous oils do not spread rapidly, do not penetrate soil as readily, and affect the ability of pumps and skimmers to handle the oil. (See also **Light ends, Volatility.**)

Volatile organic compounds (VOCs): These are organic compounds with vapor pressure high enough to cause the compounds to evaporate at normal temperatures.

Volatility: This is the tendency of a solid or liquid substance to pass into the vapor state. Many hydrocarbons with low carbon numbers are very volatile and readily pass into a vapor state when spilled. For example, gasolines contain a high proportion of volatile constituents that pose considerable short-term risk of fire or explosion when spilled. On the other hand, bunker fuels contain few volatile hydrocarbons as they are removed during the fractional distillation refining process.

Water-in-oil emulsion: This is a type of emulsion in which droplets of water are dispersed throughout oil. It is formed when water is mixed with a relatively viscous oil containing significant amounts of resins and asphaltenes, by wave action. This type of emulsion is sometimes stable and may persist for months or years after a spill. Water-in-oil emulsions containing 50% to 80% water are most common, range in consistency from greaselike to solid, and are generally referred to as "chocolate mousse." (See also **Chocolate mousse, Emulsification, Oil-in-water emulsion.**)

Water table: The fluctuating upper level of the water-saturated zone (groundwater) located below the soil surface. (See also **Groundwater.**)

Waxes: Waxes are the larger saturate compounds, carbon number 18 and up, that make up oil. They consist of long-chain organic compounds. (See also **Paraffin, Saturate group.**)

Weathering: This refers to a series of processes whereby the physical and chemical properties of oil change after a spill. These processes begin when the spill occurs and continue indefinitely while the oil remains in the environment. Major processes that contribute to weathering include evaporation, emulsification, natural dispersion, dissolution, photooxidation, sedimentation, adhesion to materials, interaction with mineral fines, microbial biodegradation, and the formation of tar balls or tar mats. (See also **Biodegradation, Dispersion, Dissolution, Emulsification, Photooxidation, Sedimentation, Tar balls or mats.**)

Weir skimmers: This is a type of skimmer that uses the force of gravity to drain oil from the water surface into a submerged holding tank. Basic components include a weir or dam, a holding tank, and an external or internal pump. As oil on the water surface falls over the weir or is forced over by currents into the holding tank, it is continuously removed by the pump.

Reading for Further Information

Chapter 1

Oil Import Data

International Petroleum Encyclopedia, Pennwell Publishing Company, Tulsa, OK, Annual.

Spill Statistics

Etkin, D.S., "Spill Occurrences: A World Overview," Chapter 2, in *Oil Spill Science and Technology*, M. Fingas, editor, Gulf Publishing Company, New York, pp. 7–48, 2011.

Oil Origins in the Sea

National Research Council, *Oil in the Sea III: Inputs, Fates, and Effects*, National Academies Press, Washington, DC, 2003.

Chapter 2

Contingency Planning

Fingas, M., "Introduction to Oil Spill Contingency Planning and Response Initiation," Chapter 28, in *Oil Spill Science and Technology*, M. Fingas, editor, Gulf Publishing Company, New York, pp. 1027–1031, 2011.

Chapter 3

Oil Chemistry

Fingas, M., "Introduction to Oil Chemistry and Properties," Chapter 3, in *Oil Spill Science and Technology*, M. Fingas, editor, Gulf Publishing Company, New York, pp. 51–59, 2011.

Oil Properties

Hollebone, B., "Measurement of Oil Physical Properties," Chapter 4, in *Oil Spill Science and Technology*, M. Fingas, editor, Gulf Publishing Company, New York, pp. 63–86, 2011.

Chapter 4

Oil Behavior

National Research Council, *Oil in the Sea III: Inputs, Fates, and Effects*, National Academies Press, Washington, DC, 2003.

Oil Spill Modeling

Fingas, M., "Introduction to Oil Spill Modeling," Chapter 8, in *Oil Spill Science and Technology*, M. Fingas, editor, Gulf Publishing Company, New York, pp. 187–200, 2011.

Chapter 5

Oil Analysis

Fingas, M., "Introduction to Oil Chemical Analysis," Chapter 5, in *Oil Spill Science and Technology*, M. Fingas, editor, Gulf Publishing Company, New York, pp. 87–109, 2011.

Oil Remote Sensing

Fingas, M., and Brown, C.E., "Oil Spill Remote Sensing: A Review," Chapter 6, in *Oil Spill Science and Technology*, M. Fingas, editor, Gulf Publishing Company, New York, pp. 111–169, 2011.

Chapters 6 to 8

General

Fingas, M., "Physical Spill Countermeasures," Chapter 12, in *Oil Spill Science and Technology*, M. Fingas, editor, Gulf Publishing Company, New York, pp. 303–337, 2011.

Effects of Weather on Booms and Skimmers

Fingas, M., "Weather Effects on Spill Countermeasures," Chapter 13, in *Oil Spill Science and Technology*, M. Fingas, editor, Gulf Publishing Company, New York, pp. 339–426, 2011.

Chapter 9

Dispersants Generally

Fingas, M., "Oil Spill Dispersants: A Technical Summary," Chapter 15, in *Oil Spill Science and Technology*, M. Fingas, editor, Gulf Publishing Company, New York, pp. 435–582, 2011.

Dispersant Application

Fingas, M., "A Practical Guide to Chemical Dispersion for Oil Spills," Chapter 16, in *Oil Spill Science and Technology*, M. Fingas, editor, Gulf Publishing Company, New York, pp. 583–610, 2011.

Surface Washing Agents

Fingas, M., and Fieldhouse, B., "Surface-Washing Agents," Chapter 21, in *Oil Spill Science and Technology*, M. Fingas, editor, Gulf Publishing Company, New York, pp. 683–711, 2011.

Treating Agents Generally

Fingas, M., "Spill-Treating Agents," Chapter 14, in *Oil Spill Science and Technology*, M. Fingas, editor, Gulf Publishing Company, New York, pp. 429–433, 2011.

Chapter 10

In-Situ Burning Generally

Fingas, M., "In-Situ Burning," Chapter 23, in *Oil Spill Science and Technology*, M. Fingas, editor, Gulf Publishing Company, New York, pp. 737–903, 2011.

Chapter 11

Shoreline Cleanup

Owens, E.H., *Field Guide for the Protection and Cleanup of Oiled Shorelines*, Environment
   Canada, Ottawa, Ontario, 2009.

Chapter 13

General Effects

National Research Council, *Oil in the Sea III: Inputs, Fates, and Effects*, National
   Academies Press, Washington, DC, 2003.
Shigenaka, G., "Effects of Oil in the Environment," Chapter 27, in *Oil Spill Science
   and Technology*, M. Fingas, editor, Gulf Publishing Company, New York,
   pp. 985–1020, 2011.

Index